MULTICULTURAL WOMEN OF SCIENCE

Three Centuries of Contributions
with Hands-On Activities and Excercises for the School Year

by

Leonard Bernstein
Alan Winkler
Linda Zierdt-Warshaw
Reviewers

Margaret W. Carruthers
AMERICAN MUSEUM OF NATURAL HISTORY
DEPT. OF EARTH AND PLANETARY SCIENCES

Margarita Lopez-Bernstein
EDUCATIONAL CONSULTANT
LATINO AND MULTICULTURAL EDUCATION

The Peoples Publishing Group, Inc.

Free to Learn, to Grow, to Change
1-800-822-1080

Editor, Charmaine Harris-Stewart
Copy Editor, Grant Hansen
Cover Design, Northeast Visual Media Ltd.
Design, Jill Woods, Doreen Smith, Judy Pines
Illustrations, Judy Pines, Brooke Kaska
Production/Electronic Design, Margaret Lepera, Doreen Smith, Kristine Liebman
Photo Research, Kristine Liebman, Daniel Ortiz

PHOTO CREDITS

pg. 4, Cold Spring Harbor Laboratory Archives; pg 8, Cold Spring Harbor Laboratory Archives; pg. 12, AP/Wide World Photos; pg. 16, Collection of Dr. Margulis; pg. 20, Special Collections, CA Academy of Sciences; pg. 24, Archive Photo; pg. 28, Courtesy of Elizabeth Swerda for Dr. Wong-Staal; pg. 32, from *Fighting For Life* by Sara Josephine Baker; Arno Press, 1974; pg. 38, AP Wide World Photos; pg. 42, *Special thanks to* Adele Logan Alexander; pg. 46, Washington University in St. Louis; pg. 50, Bettmann; pg. 54, Courtesy of Clair Holly for Dr. Elion; pg. 58, Courtesy of Susan Benton for Dr. Elders; pg. 62, Nebraska State Historical Society, Medical College of Pennsylvania; pg. 68, New York State Museum; pg. 72, The Ferdinand Hamburger, Jr. Archives of the Johns Hopkins University; pg. 76, A/P Wide World Photo; pg. 80, Environmental Science Services Administration; pg. 84, NASA; pg. 88, Harvard College Observatory; pg. 92, Harvard College Observatory; pg. 96, NASA; pg. 102, Archive Photo; pg. 106, Archive Photo; pg. 110, AP/Wide World Photos; pg. 114, Bettman; pg. 118, Archive Photo; pg. 122, Archive Photo; pg. 128, Archive Photo; pg. 129, Courtesy of Leonard Bernstein; pg. 132, The Bancroft Library, University of California Berkeley; pg. 136, A/P Wide World Photos; pg. 140, A/P Wide World Photos; pg. 144, The U.S. Patent and Trademark Office; pg. 146, Cleveland Health Sciences Library; pg. 154, Special Collection of the Telfair Museum of Art, Savannah, Georgia, James Frothingham, artist, Daniel Grantham, photographer; Cover Photos, Collection from Photo Gallery software, Key Photos software & Courtesy of Judy Rada

ISBN 1-56256-702-0

© 1996

The Peoples Publishing Group, Inc.
299 Market Street
Saddlebrook, NJ 07663

Printed in the United States of America.

10 9 8 7 6 5 4

CONTENTS

Each chapter contains a biographical sketch of a notable woman of science, a **Hands-On Activity** called **It's Your Turn**, and a **Think Work Act** page with critical thinking questions and at least four activities featuring a variety of skills and learning styles.

WOMEN OF LIFE SCIENCE

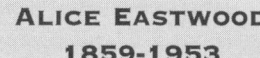

ALICE EASTWOOD
1859-1953

MARGARET MEAD
1901-1978

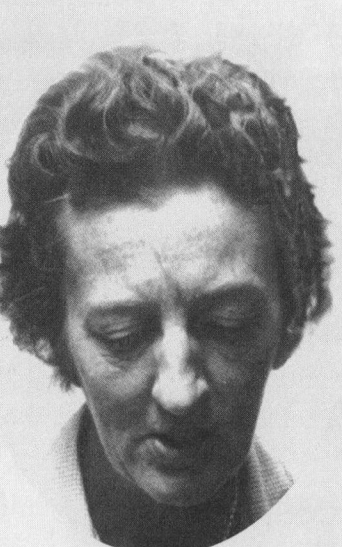

MARY LEAKEY
1913-1996

LYNN MARGULIS
1938-

SARA JOSEPHINE
BAKER
1873-1945

BARBARA
McCLINTOCK
1902-1992

ROSALIND
FRANKLIN
1920-1958

FLOSSIE WONG-STAAL
1946-

BARBARA McCLINTOCK
1902-1992

Indian corn, easily recognized by its colored kernels, is often used for decoration. In the 1920s, Barbara McClintock became curious about the color and arrangement of the kernels of Indian corn. Her curiosity led her to study the corn's genetic makeup. It also led her to achieve the highest honors in all of science.

Barbara McClintock was born in Hartford, Connecticut, in 1902. She was one of four children of Sara and Dr. Thomas H. McClintock. In 1908, the family moved to Brooklyn, New York, where Barbara attended elementary school. Later, after attending high school, Barbara went to Cornell University. In 1923, she received her BA in cytology, the study of cells. Within three years, she also earned her master's and doctoral degrees.

While working on her master's degree, McClintock became interested in the genetic properties of corn. Under the microscope, she identified individual corn chromosomes by their unique features. This discovery played a major role in her later research.

McClintock received a fellowship from the National Research Council in 1931. Over the next two years, she continued her research at Cornell, the University of Missouri, and the California Institute of Technology. In 1936, McClintock joined the faculty of the University of Missouri; however, she left in 1941 because gender discrimination kept her from advancing. She accepted a position at the Cold Spring Harbor Laboratory, a genetics research facility, in Long Island, New York.

Throughout the 1940s, McClintock earned recognition for her research. She was elected president of the Genetics Society of America in 1944. She was also named to the National Academy of Sciences—becoming only the third woman to receive this honor. During the same period, McClintock continued her research with Indian corn. She noted that in some plants, color changes from one generation to the next did not follow a predictable pattern. This led her to believe that some genes had changed their positions on the chromosomes. In 1951, McClintock concluded that the positions of genes on a chromosome are not fixed, instead, the genes sometimes move or "jump" around unpredictably.

McClintock's work was largely ignored for more than ten years. Her findings were finally confirmed in the 1960s. Today, her research on "jumping genes" is used to explain why some bacteria develop a resistance to antibiotics. Jumping genes may also explain how some normal cells change into cancerous cells. Some scientists believe jumping genes may also speed evolution.

> ### Vocabulary
> Genes are the parts of the chromosomes that carry hereditary information.

GENETICIST

For her contributions to science, McClintock received the Kimber Genetics Award from the National Academy of Sciences in 1967. In 1970, President Nixon awarded her the National Medal of Science. She also received the Albert Lasker Basic Medical Research Award—the highest science award in the United States—in 1981. In 1983, McClintock was awarded the Nobel Prize in physiology or medicine. She is only the third woman to win an unshared Nobel Prize. She is also the first woman to win the prize in physiology or medicine.

APPLICATION OF THE SCIENCE

At one time, scientists believed that the positions of genes on chromosomes were fixed. The work of Barbara McClintock proved this belief was wrong. When a gene moves or jumps to a new location on a chromosome, a mutation occurs. A mutation is the change that results when the genetic material of an organism does not make an exact copy of itself during cell division.

When a mutation occurs, the traits of an offspring differ from those of the parent. The change may benefit the organism, harm the organism, or have no observable effect on the organism. Through the process of natural selection, helpful mutations may eventually lead to the evolution of new species. In contrast, harmful mutations may result in diseases and disorders. By understanding the specific causes of certain diseases and disorders, scientists may someday be able to prevent diseases caused by mutations.

It's YOUR TURN

Barbara
McCLINTOCK

Hands-On Activity

IDENTIFYING CHANGES IN CHROMOSOMES

TABLE I: TYPES OF CHROMOSOME MUTATIONS

Duplication	Deletion	Inversion	Translocation
During replication, a chromosome recieves an extra piece from its partner and some genes are duplicated.	During replication, one or more genes are not copied and are lost entirely from the chromosome.	During replication, a piece of chromosome breaks free and reattaches itself in the reverse order.	During replication, a whole chromosome or part of a chromosome attaches to the chromosome of a different pair.

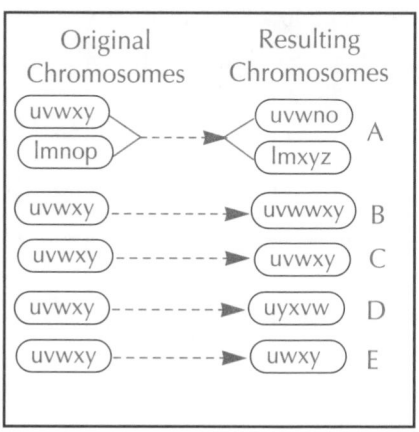

◆Each letter represents a gene on the chromosome.

BACKGROUND INFORMATION

Chromosome mutations occur when the arrangements or numbers of genes on the chromosomes change during cell division. These changes occur by duplication, deletion, inversion, and translocation. Each type of change is described in Table 1.

PROCEDURE

1. Study the four types of chromosome changes described in Table 1.
2. The figure shows chromosomes before and after replication. Use Table 1 to identify the mutations shown by the lettered chromosomes. Label the chromosome that matches the original chromosome, *Normal.*

ANALYZE AND CONCLUDE

On separate paper, write your answers.

1. Which chromosome pair did you label *Normal*? Why?
2. Which chromosome pair shows deletion? Which genes are missing from the chromosome?
3. Which chromosome pair shows duplication? Which genes are duplicated?
4. Which chromosome pair shows inversion? How do the genes of this chromosome differ from those of the original chromosome?

6

think WORK ACT

CRITICAL THINKING Answer the following questions in complete sentences.

1. Why do you think the genes studied by Barbara McClintock are called "jumping genes"?

   _____

   _____

2. Why might "jumping genes" help speed the process of evolution?

   _____

   _____

3. Barbara McClintock won her Nobel Prize in physiology or medicine. How might the genetic traits of corn be related to physiology or medicine?

   _____

   _____

GOING FURTHER Complete three of the following.

BUILD YOUR PORTFOLIO

Use a biology text to find the meanings of the terms *dominant, recessive, pure,* and *hybrid*. Find out what a Punnett square is and how it works. Then create a Punnett square to show how two organisms, each having pure genes for a particular trait, can produce an offspring that is hybrid for that trait.

PERFECT YOUR SKILL

Use a biology text to learn how a blending of traits occurs in four-o'clock flowers. Draw a Punnett square showing how pink flowers form in four-o-clock flowers. Label your Punnett square carefully.

JOURNAL WRITING

Barbara McClintock worked in the same area of genetics for almost 60 years. In your journal, write a brief description of some things you hope to accomplish over the next 60 years of your life.

RESEARCH AND REPORT

Gregor Mendel, Thomas Hunt Morgan, and Barbara McClintock have been called the "three Ms" of genetics. Research the work of these scientists and summarize their contributions to genetics in a brief essay.

COOPERATIVE LEARNING

Use pieces of macaroni to represent genes. With your group, label the pieces of macaroni and arrange them to model the changes that occur during the four kinds of chromosome mutations described in Table 1 on page 6. Glue your models to a sheet of paper. Label and write captions to explain each process.

ROSALIND FRANKLIN

1920-1958

DNA is a chemical substance that is present in all the cells of your body. This substance makes you different from every other person. The discovery of the structure of the DNA molecule, which was first reported in 1953, is considered by many scientists to be the most significant discovery of the 20th century; however, the woman whose work was instrumental in finding the structure of DNA is often overlooked when credit for the discovery is given. This woman's name is Rosalind Elsie Franklin.

Rosalind Elsie Franklin was born in London, England, in 1920. As a child, Franklin received her education at the St. Paul's School for Girls. It was one of the few girls' schools that taught physics and chemistry. Franklin enjoyed science, especially astronomy. She often used star maps from local newspapers to locate and identify constellations. A star map uses dots to indicate the position of stars and constellations in the sky. By age fifteen, Franklin decided science would be her career.

Franklin entered Newnham College, Cambridge University's women's college, in 1938. She majored in chemistry and graduated in 1941. In 1942, Franklin took a job with the British Coal Utilization Research Association. While there, she discovered how the structures of coal and carbon change when they are heated.

Between 1942 and 1946, Franklin published five papers on how to use X-rays to study the structures of coal and carbon. These writings established her as an outstanding research chemist. Much of what is known about the high-strength carbon fibers and graphite used in today's nuclear power plants is based on Franklin's research.

In 1947, Franklin moved to France to work for the Central Laboratory of Chemical Sciences in Paris. Four years later, she returned to England to study biological crystals at Kings College of the University of London. There, Franklin began working with biophysicist Maurice Wilkins on DNA. Scientists knew that DNA existed, but they did not know its purpose. Together, Franklin and Wilkins searched for the structure and function of DNA (deoxyribonucleic acid). At the same time Franklin and Wilkins were doing their work, several other scientists were also involved in DNA research. Among them were James Watson and Francis Crick, working at Cambridge University in England, and Linus Pauling, working in the United States; however, Franklin's work in the X-rays of pure DNA was superior to the others. Even now, her X-

> ### Vocabulary
> DNA is the chemical building block of all living things and is responsible for transmitting hereditary characteristics.

rays of DNA are considered to be the best ever obtained. From them, Franklin was able to create a single-stranded model of DNA that showed its atoms in their proper places. In the race to discover the structure of DNA, Franklin was in the lead.

Franklin kept her notes in a red notebook. Without getting her permission, Wilkins showed the notes to James Watson. Using this information and Franklin's X-rays, Watson and Crick figured out the double-stranded structure of the DNA molecule. They described their findings in an article sent to the British science journal, *Nature*, on March 17, 1953. Watson and Crick sent their findings to the journal eleven days before Franklin sent hers. Thus, Watson and Crick narrowly won the race to describe the structure of DNA.

Disappointed and angry, Franklin left Kings College and went to work at Birkbeck College in London. There she used X-rays to find the molecular structures of viruses. In 1958, only five years after completing her work with DNA, Rosalind Franklin died of ovarian cancer. Four years later, Watson, Crick, and Wilkins shared the Nobel Prize in physiology or medicine for their work on DNA. Franklin did not share in the award partly because Nobel Prizes are not given to a deceased person. In addition, the men who were honored never credited Franklin for her contributions to the discovery of the structure of DNA.

APPLICATION OF THE SCIENCE

Rosalind Franklin used X-rays to determine the molecular structure of various materials. This technology allows scientists to find out what substances make up matter and how the atoms and molecules of the substances are arranged. The arrangement of atoms and molecules also provides information about the strength of the material. Today, engineers often use X-rays of steel, carbon and graphite, and other substances to determine their use in construction or products. This same technique can be used to look for flaws or areas of weakness in products and building materials made from these substances.

It's YOUR TURN

Rosalind FRANKLIN

Hands-On Activity

USING INDIRECT OBSERVATIONS

TO IDENTIFY OBJECTS

MATERIALS (per group of four)

blueprint paper, small identifiable objects such as paper clips or keys, light source, sealed baby food jar containing household ammonia, clock

SAFETY Wear goggles and a laboratory coat or apron throughout activity. Perform only in a well-ventilated area. Keep caps tightly sealed on jars of ammonia when the jars are not in use. Store or dispose of the ammonia as instructed by your teacher.

BACKGROUND INFORMATION

When an object is X-rayed, the object can often be identified by its silhouette or outline. Franklin used this concept to research the structure of molecules. X-ray film, photographic film, and blueprint paper are sensitive to light. By applying chemicals to these materials, images can be made to appear in areas exposed to varying amounts of light.

PROCEDURE

1. Examine a piece of blueprint paper. Place the paper on your desk, colored side up. Place an object, such as a key or a paper clip, in the center of the paper. Place the paper and its object under a light source for about ten minutes.
2. Remove the object from the paper. Examine the paper looking for any changes.
3. Have one group member uncap the jar. Let the paper rest on top of the jar for about fifteen seconds. Remove the paper from the jar and recap the jar.
4. Repeat step 3 for each group member.
5. Observe your piece of blueprint paper noting any changes.

6. Exchange your group s papers with those of another group. Use the silhouettes formed to identify as many objects as you can.

ANALYZE AND CONCLUDE

On separate paper, write your answers.

1. What changes, if any, did you observe in the appearance of the blueprint paper between steps 1 and 5? Describe the processes you think caused these changes.
2. What objects were you able to identify from the images formed on the paper? What made it possible for you to identify these objects?
3. How are the results you obtained similar to those obtained through the use of X-rays?

10

think WORK ACT

CRITICAL THINKING Answer the following questions in complete sentences.

1. How did the structure of DNA initially described by Rosalind Franklin differ from that described by Watson and Crick?

   _____

   _____

2. Graphite and diamond are made up only of carbon atoms. Graphite is one of the softest materials known. Diamond is the hardest substance known. What do you think explains the difference in the hardnesses of the two substances?

   _____

   _____

3. What information did Franklin's work with X-rays provide to science?

   _____

GOING FURTHER Complete three of the following.

BUILD YOUR PORTFOLIO

Use a biology text or other reference book to find the structure of DNA. Make a drawing that shows the molecule's structure. Label the sugars, phosphate bases, and nitrogen pairs in the molecule. Add a caption that explains how the nitrogen pairs are arranged.

RESEARCH AND REPORT

DNA and RNA are both nucleic acids. Use library references to find out how they are alike and how they are different. Write a report of your findings.

JOURNAL WRITING

If Rosalind Franklin had lived until the Nobel Prize was awarded for the discovery of the structure of DNA, do you think she would have shared in the award? Why or why not?

COOPERATIVE LEARNING

Work in a group to construct a model that shows the structure of DNA. Before beginning your model, use a biology text or encyclopedia to see how the parts of DNA are arranged to form base pairs. Then select the materials you will need to construct your model. You may want to use colored paper, string, vegetable dye, pasta, etc. After your model is completed, display it where it can be viewed by your classmates.

CONCEPT MAPPING

Develop a concept map that identifies the scientists who were searching for the structure of DNA. Arrange your concept map to show what information was learned by each scientist and how information from different scientists was combined to discover the structure of DNA.

MARY LEAKEY
1913 -1996

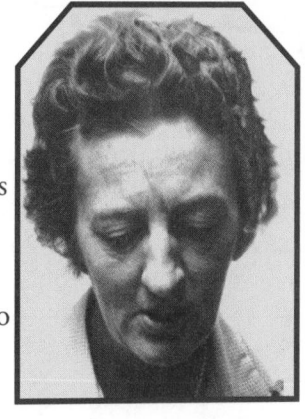

Earth is about 4.5 billion years old. Humans, however, have lived on Earth for only five to eight million years. Much of what is known about early humans comes from the work of the Leakey family (Mary Leakey, her husband Louis, and their son Richard). Working together and individually, the Leakeys uncovered fossils of several groups of early humans. All three members of the Leakey family were paleontologists.

Mary Leakey was born in London, England. She was the only child of Cecilia and Erskine Nicol. Because the family traveled a lot, Mary had no formal early education. When she was thirteen, her father died and Mary was sent to a convent school. Mary had difficulty adapting to a formal school setting. She was expelled from the only two schools she attended.

Mary had a natural talent for drawing. She also shared an interest in archaeology with her father. Archaeology is the study of ancient peoples. In her autobiography, *Disclosing the Past*, Mary speaks of trips she took with her father to study prehistoric cave paintings. She credits these trips with awakening her interest in archaeology.

Between 1930 and 1934, Mary attended lectures in geology and anthropology at University College, London, and at the London Museum. During summers, she worked as an illustrator for archaeologist Dorothy Liddell. Mary made drawings of the objects found at dig sites. These drawings brought Mary to the attention of archaeologist Gertrude Canton-Thompson. Later, Canton-Thompson introduced Mary to Louis Leakey, who asked Mary to do the drawings for his book, *Adam's Ancestors*. The book was published in 1934. Two years later, Louis and Mary married.

Mary traveled with Louis to Africa. For many years, she assisted her husband in his work and raised their children. In 1948, Mary made an important discovery. She found the ape-like skull of a forerunner to humans called *Proconsul africanus*. The discovery earned the Leakeys worldwide attention. It also helped them get funding to continue their work.

On July 17, 1959, Mary made another major discovery. Working in the Olduvai Gorge of Tanzania, she found the upper jaw and a full set of teeth of an early hominid (human). The bones were from a 1.75-million-year-old hominid called *Zinjanthropus*. The find was later reclassified as *Australopithecus boisei*. Over the next few months, Mary found more hominid bones and many stone tools.

In 1964, the Leakeys found a skull that was

> ### Vocabulary
> Paleontology is the study of the fossils of living things from long ago.

more similar to modern humans than the earlier skull. They named this hominid *Homo habilis*, a name meaning "able man."

Mary and Louis Leaky's marriage fell apart soon after this discovery. Although they did not officially separate or divorce, the Leakey's work kept Louis and Mary in different parts of the world. Louis traveled throughout the world giving lectures about the importance of the work being done by his family. Mary remained in Africa and continued her work. Much of this work was done in Laetoli, an area in East Africa's Serengeti Plain.

In 1972, Louis Leakey died of a heart attack.

Three years later, Mary discovered the 3.75-million-year-old jaws and teeth of a hominid in the Laetoli region of Tanzania. In 1979, Mary made another solo discovery in Laetoli. This time, she discovered the twenty-three-meter long trail of hominid footprints in volcanic ash. The footprints were more than 3.6-million years old. Mary believes the footprints are her most important find. Unlike earlier discoveries, this discovery showed that early hominids walked upright, as people do today.

Mary Leakey died in 1996. Her life-long contributions to paleontology have earned her recognition as an authority in human evolution.

MAJOR HOMINID ARCHAEOLOGICAL FINDS

COOPERATIVE/
COLLABORATIVE
SCIENCE

The work of Mary Leakey provides much information about early humans; however, a more complete picture of how humans have changed over time involves the work of many scientists. The table shown here summarizes some major finds that have been used to piece together the evolutionary development of humans.

Scientist	Find	Genus/Species	When Hominid Lived (millions of years ago)	Year of Find
Team in Ethiopia	teeth, jaws, skull base, arm bones	*Australopithecus ramidus*	4.4	1994
Donald Johanson	female skeleton named Lucy	*Australopithecus afarensis*	4.4 -3	1974
Raymond Dart	skull of child	*Australopithecus africanus*	2.7-2.1	1924
Mary Leakey	upper jaw and teeth	*Australopithecus bosei*	2-1	1959
Mary & Louis Leakey	skull and stone tools	*Homo habilis*	3.7-1.5	1964
Richard Leakey	complete skeleton	*Homo erectus*	1.5	1984

Multicultural Women of Science ◆ Mary Leakey

It's YOUR TURN

Mary LEAKEY

Hands-On Activity

MODELING HOMINID BRAIN SIZES

MATERIALS (per group of four)

4 balloons of different colors, 100-mL graduated cylinder, sink or pail

SAFETY

Clean up any spills that occur immediately.

BACKGROUND INFORMATION

From their studies of hominid skulls, paleontologists have observed that cranial capacity has changed over time. Cranial capacity is a measure of the size of the part of the skull that holds the brain. Using data about cranial capacity, scientists have estimated the brain sizes of various hominids by volume (in cm^3).

PROCEDURE

1. Select a hominid from the table below. (Each group member should select a different hominid.)

HOMINIDS

Hominid	Cranial Capacity (in cm³)
Homo sapiens (Modern humans)	1400
Homo habilis	750
Australopeihiecus africanus	500
Homo erectus	1000

2. Select a balloon. Use the graduated cylinder to add enough water to the balloon to create a brain of the correct size for your hominid. Note: 1 mL of water = 1 cm^3. Tie off the neck of your filled balloon.

3. Arrange the brains in order from smallest to largest.

ANALYZE AND CONCLUDE

On separate paper, write your answers.

1. How many mL of water were added to each balloon to create a brain of the correct size? Why?

2. Using the sequence of balloons you created, identify and name the sequence of hominids from smallest to largest brain size.

3. Use your answer from question 2 to summarize the change in brain size in humans over time.

4. What relationship, if any, do you think there is between intelligence and brain size in humans? Explain your answer.

14

think WORK ACT

Mary
LEAKEY

CRITICAL THINKING Answer the following questions in complete sentences.

1. The oldest fossils of humans have been discovered in Africa. What does this information suggest?

_____

_____

2. Stone tools were found near the bones of *Homo habilis*. What does this finding suggest about this early human species?

_____

_____

3. What observations may have led Mary Leakey to conclude that the hominid that left the 3.6 million-year-old footprints walked upright?

_____

_____

GOING FURTHER Complete three of the following.

BUILD YOUR PORTFOLIO

Use the table on p.13 to create a timeline that shows when each hominid species lived.

ALTERNATIVE ASSESSMENT

Design a way to illustrate how the footprints of a hominid that walked on two legs would differ from those of a hominid that walked on both hands and feet. Demonstrate your findings to the class.

RESEARCH AND REPORT

Neanderthals and Cro-Magnons were two early groups of modern humans *(Homo sapiens)*. Do research to find out when these two groups of *Homo sapiens* lived and how they lived. Present your findings in an oral report.

COOPERATIVE LEARNING

Find out what kinds of tools were used by *Homo habilis* and the *Homo sapiens* known as the Neanderthals and Cro-Magnons. Work with your group to make models of these tools. Display the models and demonstrate how they may have been used.

JOURNAL WRITING

Imagine you lived at the time of *Homo habilis*. How would your life be different than it is today? Conduct additional research to support your answer.

PERFECT YOUR SKILL

Use library resources to find out where in the world each hominid find listed in the table on p.13 was made. Plot the location of each find on a map. Include a key for the map.

LYNN MARGULIS
1938-

As you read this sentence, bacteria called *E. coli* are thriving in your small intestine. These bacteria help you digest food. You provide the bacteria with a home that meets all their needs. This type of relationship, in which two different organisms live in a close association, is called *symbiosis.* Biologist Lynn Margulis has used the idea of symbiosis to explain how cells that lack nuclei and other organelles may have evolved into more complex cells.

Lynn Margulis was born Lynn Alexander in Chicago, Illinois. As a young girl, she loved to read and write. She also had several part-time jobs, both to earn money and to occupy her spare time. At the end of her sophomore year of high school, fifteen-year-old Lynn applied for an "early entry" degree program at the University of Chicago. She was accepted.

Students taking science courses at the university were asked to read original works in science. These readings fascinated and inspired Lynn. She decided that science research would become her life's work. While in school, Alexander met Carl Sagan, a physics graduate

> ## Vocabulary
> **Mitochondria** are cell organelles that carry out respiration and provide energy to the cell.
> A **chloroplast** is the cell structure that contains chlorophyll; usually found in plant cells.

student who would later become a noted astronomer. After she received her bachelor's degree in 1957, Alexander and Sagan married.

The Sagans moved to Wisconsin, where Lynn enrolled in a graduate program at the University of Wisconsin to study zoology and genetics. After she received her master's degree in 1960, the Sagans moved to California.

Lynn Sagan began research in genetics at the University of California, at Berkeley. In 1965, she received her Ph.D. That same year, she and Carl Sagan moved to Massachusetts with their two sons. Shortly after, the Sagans were divorced.

Margulis worked for a short time developing teaching materials at Brandeis University in Waltham, Massachusetts, and at ESI, Elementary Sciences Studies in Watertown, Massachusetts. She accepted a position at Boston University. For the next twenty-two years, she worked in the university's biology department. During this time she married Thomas Margulis, a crystallographer, and had two more children.

Boston University proved to be the safe haven Lynn Margulis had been searching for. She produced some of her best work there. Using biochemical methods and the electron microscope, she and many others showed that DNA, the material that carries genetic information, was also present outside the cell nucleus. This DNA was located in mitochondria and chloroplasts, and differed from the DNA in the nucleus. Further research showed that the organelles containing the DNA resembled bacteria. From these data, Margulis formed ideas about how cells that have nuclei evolved through bacterial partnerships. These ideas are known as the endosymbiotic hypothesis.

The endosymbiotic hypothesis states that certain cell structures, such as mitochondria, once existed as bacteria outside cells. These cells were taken in, probably as food, by larger cells, but were not digested. After fighting for survival the smaller cells became a working part of the larger cell. When the larger cell reproduced, the smaller cells inside were also reproduced. Over time, a permanent relationship, in which the cell and the former bacteria worked as a unit, was established.

Lynn Margulis first suggested her hypothesis in 1966. At the time, most scientists did not think the hypothesis had much value; however, many scientists have changed their views. Today, the hypothesis is often used to explain how simple organisms may have evolved into more complex organisms. Since first stating the endosymbiotic hypothesis, Margulis has written many articles and books to explain her biological research. She has written five books and more than a dozen articles with her son, Dorian Sagan. Since 1988, Margulis has worked as a professor in the biology department of the University of Massachusetts at Amherst.

COOPERATIVE/ COLLABORATIVE SCIENCE

Lynn Margulis's research and hypotheses are related to and supported by ideas in other areas of science, such as genetics. One mystery she tried to solve was how the properties of a cell might differ from the cell from which it had formed. Margulis suggested that such differences might result from genes in other parts of the cell, possibly the cytoplasm. Cytoplasm is material outside the cell nucleus.

Other researchers were involved in similar studies and, with Margulis, showed that genes are present in organelles outside the nucleus. These organelles include the chloroplasts of plant cells and the mitochondria of plant and animal cells. Using the electron microscope, Hans Ris, a former professor of Lynn Margulis s, found a close resemblance between chloroplasts and certain kinds of bacteria. Additional studies by Margulis and her son, Dorian Sagan, showed that the reproductive process in mitochondria is similar to that of bacteria.

In 1989, the research of Kwang Jeon of the University of New York at Buffalo showed that a certain type of bacteria attacked the amoebas he was studying. Many of the amoebas died, but some organisms survived even though their cytoplasm contained many bacteria. Over time, some new generations of amoeba were unable to survive without the bacteria in their cytoplasm. Similarly, the bacteria could not survive without the amoebas. Such evidence helps to support the ideas Margulis stated in her endosymbiotic hypothesis.

It's YOUR TURN

Lynn MARGULIS

Hands-On Activity

LOOKING AT SYMBIOTIC RELATIONSHIPS

MATERIALS (per group of two)

lichen sample, forceps, microscope slide, baby-food jar containing water, medicine dropper, dissecting needle, cover slip, compound microscope

SAFETY

Work carefully with the dissecting needle, microscope slide, and cover slip to avoid cutting yourself.

BACKGROUND INFORMATION

A symbiotic relationship is one in which two different organisms live closely together to the benefit of at least one of the organisms. An example is the relationship between the crocodile bird and the crocodile. The crocodile bird picks at and removes bits of meat from the teeth of the crocodile. In return for this service, the crocodile does not harm the bird, and the bird gets a meal.

Another symbiotic relationship exists between the algal and fungal cells that make up an organism known as a lichen. In a lichen, the algal cells are usually fixed within the tangled, thread-like mass of the fungus. Through photosynthesis, the alga provides nutrients for the fungus. In return, the fungus provides the alga with water, minerals, and protection against the elements.

PROCEDURE

1. Use the forceps to place the lichen sample on a clean microscope slide.
2. Add two drops of water to your sample.
3. Hold the sample in place with your forceps. Gently pull your sample apart using the dissecting needle. Add a cover slip.
4. Using low power, examine your sample under a microscope. Draw what you see on a sheet of paper.

ANALYZE AND CONCLUDE

On separate paper, write your answers.
1. Identify and label the alga in your drawing.
2. Identify and label the fungus in your drawing.
3. What is the job of algae in lichens?
4. What is the job of fungi in lichens?
5. Explain how the alga and the fungus in a lichen depend upon each other for survival.
6. What type of relationship exists between the alga and fungus in a lichen?

Name _____ Date _____

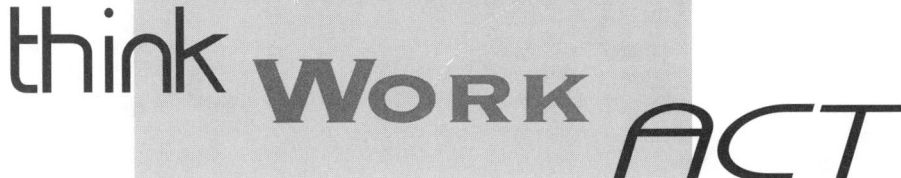 

CRITICAL THINKING Answer the following questions in complete sentences.

1. What is symbiosis?

   _____

   _____

2. In a parasitic relationship, one organism called a parasite lives on or in another organism called the host. The parasite obtains its food from the host. The host is harmed, but not usually killed, in the relationship. Is parasitism an example of symbiosis? Why or why not?

   _____

   _____

3. How does the endosymbiotic hypothesis of Lynn Margulis relate to evolution?

   _____

   _____

GOING FURTHER Complete three of the following.

BUILD YOUR PORTFOLIO

Look for examples of symbiosis between organisms living in your area. Photograph or make drawings of three examples you locate. Write a caption for each photograph or drawing explaining how each organism is affected by the relationship.

PERFECT YOUR SKILL

Look up the meanings of the terms *predation, commensalism,* and *mutualism* in a dictionary. Explain which of the processes described by the terms are examples of symbiosis.

JOURNAL WRITING

In your journal, describe a relationship from your life that involves symbiosis. Explain how each participant is involved in the relationship.

CONCEPT MAPPING

Many relationships exist between and among organisms. Some of these relationships involve predation, commensalism, parasitism, mutualism, competition, and neutralism. Find the meaning of each term. Create a concept map to show how the terms are related.

ALICE EASTWOOD

1859-1953

Rosaceae is the scientific name for the rose family. Scientists all over the world give a Latin name to every organism they discover. By doing this, an organism has the same name in every country of the world no matter what language is normally spoken by its people. At age six, Alice Eastwood began learning the Latin names of many plants by reading books given to her by an uncle.

Alice Eastwood was born in Toronto, Canada, in 1859. She was one of three children born to Colin Skinner Eastwood and Eliza Jane Gowdey. When Alice was six, her mother died and the children were sent to live with their uncle, Dr. William Eastwood, on his country estate.

While living with her uncle, Alice was encouraged to learn about plants; however, she lived with her uncle only a short time. She and her sister were soon sent to the Oshawa Convent in Toronto, Canada. While living at the convent, Alice's interest in plants continued. She met a gardener who taught her many gardening skills which she used in her future work with plants.

After six years at the convent, Alice was reunited with her family. She and her sister

Vocabulary

An herbarium is a collection of dried plants that is classified and displayed for botanical study.

went to live with her father and brother in Colorado. There Alice entered the East Denver High School, eventually graduating as class valedictorian in 1879. She spent the next ten years teaching at her former high school.

While Eastwood worked as a teacher, her interest in plants continued. During summer vacations, she explored the Rocky Mountains collecting plants and making drawings of them. In 1890, Eastwood left her teaching job to continue her study of plants. For the next two years, she traveled between California and Colorado, studying the plants in remote areas of both states. In 1892, Eastwood took a job with the California Academy of Sciences in San Francisco. She was hired to work as an assistant to Katherine Brandages, who was the curator of the academy's herbarium. The following year, Eastwood published her book, *A Popular Flora of Denver, Colorado*. The book included the data Eastwood gathered during her plant explorations in Colorado.

After Eastwood was with the academy for two years, Brandages left the academy. For the next fourteen years, Eastwood organized and enlarged the academy's plant collection at the herbarium. While organizing the plants, Eastwood decided to group the academy's rare plants together in an area of the academy that was separate from the main collection. The separation from the main collection proved invaluable in 1906. In that year, San Francisco was struck by a severe earthquake. The earthquake and the fires it caused destroyed the academy. Luckily, the rare plants were located in one area, so most of this collection was rescued before the academy was destroyed.

Eastwood spent the next six years working to rebuild the rest of the academy's collection. She made many field trips to flower and tree gardens throughout the United States to collect plants for the academy. To make sure she classified her specimens correctly, Eastwood compared the plants she gathered with those that were already classified. Eastwood also traveled to Europe to study the plants in the British Museum, the Royal Botanic Gardens, Kew Gardens of England, and the Natural History Museum of Paris.

By 1949, forty-three years after the disastrous earthquake, Eastwood had added more than 340,000 specimens to the rebuilt herbarium. She also wrote more than 300 articles and books on plants from different parts of the country. Many of her writings were illustrated with her drawings.

Eastwood's work inspired other female botanical explorers. Among them were Louise Boyd and Ynes Mexia. During her life, Eastwood received many honors, including the naming of several plant species in her honor. One of these species is a daisy called Eastwoodia. When she retired from the academy in 1949, Eastwood was asked to serve as honorary president of the International Botanical Congress in Sweden. She remained active with this group until her death at age ninety-four.

APPLICATION OF THE SCIENCE

Botanical gardens and arboretums have vast collections of plants. Often, these collections are assembled for their beauty. Plants, however, have many important uses: as sources of foods; as fibers for making textiles, paper, clothing, and other products; and as medicines. Collecting and preserving plants is a necessity because of these many uses.

In many parts of the world, plant species are preserved through seed banks. Seed banks are places in which seeds and the plant parts from which new plants can be grown are stored. The seeds and plant parts are sorted according to their genetic traits. Using the resources of seed banks, it is possible to introduce new traits into crops when the need arises. For example, if a certain variety of corn becomes threatened by a particular disease, scientists can breed the corn with another type that resists the disease. In this way, farmers can grow a healthier corn crop. Seed banks may also prevent extinction of some plant species. For example, if a plant species in a certain part of the world is destroyed by natural or human causes, it may be possible to save the species from extinction by replanting an area using plants reproduced from specimens in a seed bank.

It's YOUR TURN

Alice
EASTWOOD

Hands-On Activity

CLASSIFYING PLANTS AS MONOCOTS AND DICOTS

MATERIALS (per individual)

Pencil (colored optional), notebook

BACKGROUND INFORMATION

Most of the plants you are familiar with are angiosperms plants that reproduce from seeds formed in flowers. These are divided into two large groups monocots and dicots. Monocots form from seeds with only one seed leaf. Dicots form from seeds with two seed leaves. Other traits that help classify plants as monocots or dicots are shown in Figure 1.

PROCEDURE

1. Go to the area identified by your teacher. Make detailed sketches of the leaves and flowers (if present) of ten different plants (including grasses and trees). In your leaf sketches, show the appearance of the veins. Beside your flower sketches, write the number of petals the flower has. Number your drawings 1-10.

2. On a sheet of paper, make a data table with the column heads shown below. Include ten rows in your table.

Sample Data Table:

Drawing #	Leaf Pattern	# Petals on Flower	Monocot or Dicot?

3. Use your drawings to fill in the first three columns of your table.

4. Use Figure 1 to classify each plant as a *monocot* or a *dicot*. Write this information in the last column of your table.

SAFETY

Wear appropriate clothing when working in the field. Avoid contact with plants and insects that can cause allergic reactions. Your teacher can point these out.

Figure 1 —Monocot and Dicot Plants

Monocot Traits	Dicot Traits

Seed has one seed leaf	Seed has two seed leaves

Leaf veins usually netlike or branched	Leaf veins usually parallel

Petals and flower parts are usually in multiples of three (3, 6, 9, etc.)	Petals and flower parts are usually in multiples of four (4, 8, 12, etc.) or five (5, 10, etc.)

Roots are usually fibrous (many tiny roots)	A taproot (thick, root part with smaller roots branching from it) is usually present

ANALYZE AND CONCLUDE

On separate paper, write your answers.

1. What traits did you use to classify your plants as monocots or dicots?

2. What other traits could you use to classify a plant as a monocot or a dicot?

3. From your data, which plants were more common monocots or dicots?

4. Compare your data with the data of three classmates. How do the data compare?

think WORK ACT

CRITICAL THINKING Answer the following questions in complete sentences.

1. How does giving organisms Latin names make communication between scientists in different parts of the world easier?

   _____

   _____

2. In her book, A Popular Flora of Denver Colorado, Alice Eastwood described the different kinds of plants growing in a specific area. How might such a book be useful?

   _____

   _____

3. Do you think Alice Eastwood's informal training in botany helped or hurt her career? Could Alice Eastwood achieve similar success today?

   _____

GOING FURTHER Complete three of the following.

BUILD YOUR PORTFOLIO

Begin a leaf collection for your portfolio. Preserve leaves that have fallen to the ground by ironing them between two pieces of wax paper. Use proper safety precautions as you use the iron.

ALTERNATIVE ASSESSMENT

Obtain seeds from 10 different plants. Carefully cut open the seeds to observe the seed leaf or leaves. Glue the open seeds to a sheet of paper. Classify each seed as a monocot or a dicot.

COMMUNITY RESOURCES

Consult your state wildlife agency to find information on 10 plants from your state that are now extinct, endangered, or threatened. Find the cause of the extinction or declining numbers of the species. Make a table of the information you gather.

COOPERATIVE LEARNING

Create a field guide to the plants in your area. To make a field guide, have two or three members of the group supply photographs or drawings as well as data on where each plant was found. One or two group members should use library sources to identify each plant by its common and scientific names. The last group member should combine all the data in a notebook that will serve as the field guide.

JOURNAL WRITING

In your journal, describe three ways plants affect your everyday life. Explain how your life would be different without plants.

MARGARET MEAD
1901—1978

In 1925, a young college graduate student visited Samoa, a group of small islands located east of Australia in the central Pacific Ocean. During her six-month stay on the island of Tau, she lived among the native people, studying their culture and their teenage children. She returned to the United States in 1926 and in 1928 published her first of many books, Coming of Age in Samoa. The book created much controversy at the time. It also launched Margaret Mead's career as one of the leading anthropologists of the 20th Century.

Margaret Mead was born in Philadelphia, Pennsylvania in 1901. She was raised in a home that encouraged curiosity and education. Her father was a professor of economics at the University of Pennsylvania. Her mother was a sociologist and a strong supporter of women's rights. Her grandmother, Martha Ramsey Mead, was a child psychologist who lived with the family. Grandmother Mead taught young Margaret botany and algebra. She also taught her grandaughter to observe young children and to take notes on their behavior. This skill helped Margaret later in her career.

Margaret completed her secondary school education and enrolled at her mother's alma mater, DePauw University in Indiana as an English major. In 1920, Margaret moved to New York City and transferred to Barnard College, Columbia University, where she majored in psychology. She graduated with honors, receiving her bachelor's degreee in 1923. That same year, Mead married Luther Cressman, a divinity student. Mead remained on campus for one more year to complete her master's in psychology.

In 1925, Mead and her husband conducted a six-month field study on the behavior of adolescent girls in the Samoan Islands. To ensure her acceptance among the native people, Mead learned their language and practiced the native customs of sitting cross-legged on a pebbly floor and eating from a woven plate with her fingers. Mead was the first anthropologist to study a native people in this manner.

Mead reported her findings in her book, Coming of Age in Samoa. In this book, she suggested that children developed their personalities based on the culture in which they were raised. This idea created much controversy because it differed from the common belief that personality traits were inherited.

Mead returned to New York City in 1926. She was appointed assistant curator of ethnology (a branch of anthropology that studies different cultures) at the American Museum of Natural History. Although she continued to travel and study people of various cultures, Mead remained with the American Museum of Natural History throughout her career.

> ## Vocabulary
> **Anthropology** is the area of science that deals with the physical, social, and cultural customs of a people.

ANTHROPOLOGIST

In 1928, Mead divorced Cressman and married Dr. Reo Fortune, an anthropologist from New Zealand. That same year, she and Fortune traveled to the Admiralty Islands, north of New Guinea in the west Pacific Ocean. Between 1928 and 1929, Mead studied the children of the Manus, a group of people who live in homes built upon stilts in the water. Her study resulted in her second book, Growing up in New Guinea, published in 1930.

In 1929, Mead returned to the United States, received her Ph.D. in anthropology from Columbia University, and began her third field study. Between 1929 and 1933, Mead and Fortune studied a group of Native Americans and three groups from New Guinea. Mead and Fortune used photography to captures details of behavior that escaped their note taking.

Mead and Fortune divorced in 1935. The following year, Mead married Gregory Bateson, a British anthropologist she had met in New Guinea. Mead and Bateson traveled to Bali, an island in Indonesia, to begin a study of Balinese culture. Between 1936 and 1939, Mead and Bateson made an extensive photographic study of the Balinese people, taking nearly 40,000 photos.

During her career, Mead was an influential member of several organizations, including the American Anthropological Association, the American Academy of Arts and Sciences, the American Association for the Advancement of Science, and the National Academy of Sciences. She was also a lecturer and guest professor at various colleges.

Mead received many awards. Among them are the National Achievement Award, the Women's Geography Medal, and the Kaling Prize from UNESCO and the government of India. In 1976, Mead was inducted into the Women's Hall of Fame.

BUILDING ON THE PAST

In recent years, many nations have begun to explore the importance of cultural awareness. In addition to traveling to different parts of the world, much of what is known about peoples of different cultures has resulted from the work of many anthropologists. For example, Margaret Mead spent most of her professional life studying the cultures of Pacific Island peoples and Native Americans. Similarly, African American anthropologist Zora Neale Hurston conducted extensive research into the culture of peoples living in the South. Similar studies have been conducted in South America, Mexico, and the Aleutian Islands. In the United States today, the importance of valuing different peoples and their cultures helps develop an appreciation for these differences and builds on the belief that there is strength in diversity.

It's YOUR TURN

Margaret MEAD

FOODS OF DIFFERENT CULTURES

MATERIALS (per student)

paper and pencil, access to a reference library

BACKGROUND INFORMATION

Anthropologists study different cultures. The communication, religious ceremonies and rituals, methods of obtaining food, types of shelters, and the celebration of certain holidays and events are all concerns of anthropologists. Often, these types of activities are unique to a specific culture. The study of cultures gives us an understanding of the behavior of others as well as our own behavior. It also helps people learn how to better interact with peoples from other parts of the world.

PROCEDURE

1. Obtain cookbooks or recipes for foods eaten by peoples of a culture different from your own.
2. Choose two recipes from a specific culture. Copy the list of ingredients and the cooking instructions for each recipe.
3. Identify the country or region of the world from which the recipe originates. Decide if the foods for which you have chosen recipes should be classified as breads, appetizers, soups, entrees, beverages, or desserts.
4. Conduct library research to find out why the type of food you chose is common to a country or region of the world. For example, is seafood common because the people live on an island? Do the people living in an area have beliefs that prevent them from eating certain types of foods?
5. Write a summary of your findings. Be prepared to share your recipes and the information you learned about the foods you chose with the class.
6. Compare the foods you selected with those selected by classmates representing other cultures.

ANALYZE AND CONCLUDE

On separate paper, write your answers.

1. How did you determine the region of the world from which your recipe originated?
2. How did you decide on a classification group for your food?
3. What was the main ingredient for each of your recipes?
4. When you compared your recipes to those collected by other students, what similarities of ingredients did you find? How might you account for these similarities?
5. What types of information might you obtain from preparing and tasting foods from different cultures?

think WORK ACT

CRITICAL THINKING Answer the following questions in complete sentences.

1. Why is the work of anthropologists important?

   _____

   _____

2. What types of information are studied by anthropologists?

   _____

   _____

3. How does the work of anthropologists differ from the work of scientists working in many other areas of science?

   _____

   _____

GOING FURTHER Complete three of the following.

BUILD YOUR PORTFOLIO

Obtain pictures from magazines showing people of different cultures in traditional clothing. Find out the historical significance of different articles of clothing as well as the use of color. Summarize your findings as captions for your pictures.

ALTERNATIVE ASSESSMENT

Plan a multicultural dinner using the recipes you gathered. Prepare traditional meals representative of various cultures.

JOURNAL WRITING

Imagine you are an anthropologist. In what part of the world would you most like to conduct your research? Why?

COOPERATIVE LEARNING

Do library research to find 10 dates that are celebrated as holidays by peoples of various cultures. Circle the dates on a calendar. For each date, include a brief description of the holiday and its significance.

RESEARCH AND REPORT

Do library research to find out how foods that are common in different nations are related to geographic location. Prepare a table that explains your findings.

FLOSSIE WONG-STAAL

1946-

In some cultures, names are selected based upon historic events or a person's physical features. A young Chinese girl, Yee-Ching, received her English name when she entered school in British-controlled Hong Kong. When her father enrolled his daughter in a Catholic school, the English-speaking nuns at the school insisted that the girl be given an English name. The father was given a list of the names of typhoons that had struck Hong Kong. From this list, Yee-Ching's father selected the name *Flossie.*

Flossie Wong was born in Canton, China, in 1946. She was the youngest of four children born to Sueh-fung Wong, a cloth dealer, and his wife, Wei-Chung. In 1952, the Wongs left communist China to make their home in British Hong Kong. It was here that Flossie attended school. When Flossie entered high school, she was assigned an area of study—science. She viewed this as a privilege and did well in her studies.

In 1965, Flossie Wong came to the United States to attend college. She entered the University of California at Los Angeles. Among the courses she most enjoyed were those in molecular biology.

> ## Vocabulary
>
> Molecular Biology is the study of the structure and function of chemicals in living things.

Flossie married her classmate, Steven Staal, in 1971. The following year, she earned her Ph.D. and was named Outstanding Woman Graduate of the Year. Steven Staal went to work for the National Institutes of Health (NIH) in Maryland. In 1973, Flossie Wong-Staal also found work at the NIH. She joined the team of Robert Gallo, who was researching viruses that cause cancer.

During the 1970s, Gallo, Wong-Staal, and others worked to isolate a kind of virus, called a retrovirus, that could cause cancer in humans. In 1981, they were successful in isolating the virus from human leukemia cells. They named the virus HTLV. Gallo and his research team identified how HTLV attacks the immune system. They also showed that the virus could spread from one person to another through sexual contact, blood transfusion, and from mother to unborn child.

In the early 1980s, acquired immune deficiency syndrome (AIDS) gained national attention as a devastating disease. Gallo questioned whether HTLV or a similar virus could be the cause of AIDS. In 1984, nearly ten years after beginning its research, Gallo's team had isolated the human immunodeficiency virus (HIV), commonly called the AIDS virus. At about the same time, researchers at the Pasteur Institute in France isolated a similar virus.

With the AIDS virus isolated, Flossie Wong-Staal put her knowledge of molecular biology to work. In time, she identified the virus's structures. She was also the first scientist to successfully clone, or duplicate, the virus in amounts great enough to study. This second achievement is considered a landmark in AIDS research. Wong-Staal discovered how the virus regulates its growth and that of the cells it attacks. She also discovered that parts of the virus frequently mutate, or change. The virus's ability to mutate makes it difficult to discover a vaccine to prevent HIV infection.

In 1990, Dr. Wong-Staal became director of the AIDS research facility at the University of California at San Diego. Her goal is to develop an AIDS vaccine. She is also searching for new ways to treat the disease.

PERSPECTIVES

AIDS has resulted in the deaths of several hundred thousand people worldwide. Many times this number of people are infected with the virus that caused this disease. Because an effective treatment or cure for AIDS has not yet been found, these people are uncertain about their futures.

Although a cure for AIDS has not been discovered, many people have worked to learn more about this disease. Over the past two decades, knowledge about how AIDS is transmitted, how it effects the body, and what drugs may help to treat the effects of the disease has increased. The timeline lists some of the major developments that have led to increased awareness of AIDS.

TIMELINE

1977 Danish physician Margrethe P. Rask of Zaire dies from a rare form of pneumonia that later becomes associated with AIDS. Doctors also observe that Margrethe lacks the white blood cells (T-cells) that help defend the body against disease.

1980 Early in the year, several homosexual men in New York City develop a rare form of skin cancer called *Kaposi's sarcoma*. The men also lack vital T-cells. Several large cities report many different, but rare infections in homosexual men.

1981 The Centers for Disease Control (CDC) publish the first documents related to the unusual diseases observed in the homosexual population. GRID (Gay Related Immune Deficiency) is used to refer to the disease.

1982 Symptoms associated with GRID are observed in het-erosexual people. The name for this group of diseases is changed from GRID to AIDS (Acquired Immune Deficiency Syndrome).

1984 Scientists at the Pasteur Institute in France discover the virus (HIV) that is thought to cause AIDS. The same year Flossie Wong-Staal uses the genetic material of HIV to make copies of the virus for future study.

1985 Use of the drug AZT for people with AIDS begins. Although the drug does not cure the disease, it does seem to help prolong the lives of people with AIDS.

1996 On January 30, researchers announce that protease inhibitors used to block reproduction of viruses are shown to be successful for some AIDS patients for up to a year; however, long-term testing is still needed.

It's YOUR TURN

Hands-On Activity

ANALYSIS OF REPORTED CASES OF AIDS IN THE USA

MATERIALS (copy of the table below for each student)

AIDS CASES in the United States 1981-1993	
Year	Number of Reported AIDS Cases
1981	199
1982	744
1983	2177
1984	4446
1985	8249
1986	13,166
1987	21,070
1988	31,001
1989	33,722
1990	41,595
1991	43,672
1992	45,472
1993	88,000+*

* Data shown only for January through September

BACKGROUND INFORMATION

According to a report issued by Antonia Novello, former Surgeon General of the United States, Infection with human immunodeficiency virus (HIV), the virus that causes acquired immune deficiency syndrom (AIDS), is one of our country s greatest health challenges. The number of AIDS cases reported each year has been increasing at an alarming rate. In 1993, the number of reported AIDS cases for the period between January and September of that year exceeded 88,000 nearly double the number of reported AIDS cases for all of 1992.

PROCEDURE

1. Carefully study the table above. Use the table to make a line graph of the data.
2. Use your graph to answer the Analyze and Conclude questions.

ANALYZE AND CONCLUDE

On separate paper, write your answers.

1. What information does your graph show?
2. In what year does the fewest number of reported cases appear? The most cases?
3. Approximately how many cases of AIDS were reported in the United States in 1988?
4. During which year was the number of AIDS cases reported to be about 40,000 individuals?
5. From the data, what can you conclude about the number of AIDS cases reported annually?
6. Would you expect the actual number of AIDS cases in the United States to be higher or lower than those listed in the table? Explain your answer.
7. Based on the trend shown in your graph, would you expect the number of reported AIDS cases in the United States to increase or decrease from the latest number indicated for the year 2000? Explain.

think WORK ACT

CRITICAL THINKING   Answer the following questions in complete sentences.

1. What contributions has Dr. Flossie Wong-Staal made to AIDS research? Why are these contributions important?

    _____

    _____

2. Clones are organisms or cells formed during asexual reproduction that are genetically identical to the organism or cell from which they formed. Why would it be important to clone a virus before studying it?

    _____

    _____

3. Why would the ability of a virus to change or mutate make it difficult to develop a vaccine against the virus?

    _____

GOING FURTHER   Complete three of the following.

BUILD YOUR PORTFOLIO

Prepare a script with questions you would ask Flossie Wong-Staal if you were going to interview her about her work. Include at least ten questions to which you would like answers. Include your list of questions in your portfolio.

PERFECT YOUR SKILL

Use a biology text or other reference book to find out the stages involved in the replication of a retrovirus. Prepare a drawing or a model that shows and explains the stages in the replication process.

JOURNAL WRITING

Imagine that a cure for AIDS has been found. In your journal, write an account of the importance of this event as it might appear in a newspaper.

COOPERATIVE LEARNING

Make all the needed plans to hold an AIDS awareness day at your school. Do research about the disease and its transmission. Prepare brochures that explain what AIDS is and how people can avoid getting it. Make the arrangements needed for speakers or people working on AIDS research and education to make presentations.

RESEARCH AND REPORT

Conduct library research to find several articles written about AIDS and its transmission for each year from 1983 to present. Write a summary explaining what progress has been made in the study of this disease and how people's knowledge and understanding of the disease has changed.

SARA JOSEPHINE BAKER
1873-1945

Have you ever gotten sick while at school? If so, you were probably sent to see the school nurse or doctor. Public schools have not always had a nurse or a doctor on staff. This practice began in New York City because of the efforts of Sara Josephine Baker.

Sara Josephine Baker was born in Poughkeepsie, New York, in 1873. She was one of four children. Her father, Orlando Baker, was an attorney. Her mother, whose family helped establish Harvard College, was Jenny Harwood Baker. When Sara was sixteen, her father and brother died. Ironically, her brother died of typhoid fever, the disease that would someday earn Sara her fame.

In her youth, Sara attended Misses Thomas' Private School for Girls. Later, although discouraged by family and friends, Sara decided to study medicine. She began her medical studies at the Women's College of the New York Infirmary for Women and Children in 1894. She received her degree in 1898, graduating second in a class of eighteen. Her specialty was pediatrics.

In 1901, seeking to supplement her income from private practice, Dr. Baker became a medical inspector for the New York City Health Department. She worked mostly with immigrant families. Dr. Baker was shocked by the dangerous health and sanitary conditions in which many of the families she served lived.

Dr. Baker became the assistant commissioner of health in New York City in 1907. It was during this time that she received national attention. Working with others, Dr. Baker identified a restaurant worker named Mary Mallon as being responsible for spreading typhoid fever. Mallon, who came to be known as "Typhoid Mary," worked as a cook in a Long Island restaurant. When health officials pursued Mallon, she fled to Manhattan, and then to the Bronx. While in the Bronx, Typhoid Mary was captured. She was placed in an isolation unit to prevent further outbreaks of the disease.

In 1910, on appeal, the United States Supreme Court released Mary Mallon on the condition that she never work again. In 1914, a typhoid epidemic broke out at a mental health center in New Jersey. A second outbreak occurred at a hospital in New York. Mallon was again captured when it was discovered she had worked as a cook at both places. Mallon was placed back in an isolation unit. She remained there until her death in 1938.

Mary Mallon was immune to the typhoid bacillus; however, she was directly responsible for at least fifty-one cases of typhoid, including three deaths. Some believed that she may have indirectly caused thousands of cases of the disease.

Vocabulary

Bacteria are one-celled organisms with no true nuclei in their cells; bacteria are often classified as three groups—cocci, bacilli, or spirilla, based on shape.

PHYSICIAN

Baker's part in identifying Typhoid Mary gained her national recognition. Her early experiences with poor, immigrant families, however, led her to focus on treating children and preventing childhood diseases. She organized nurses to go into the community to meet with mothers. The nurses stressed the importance of regular bathing and adequate ventilation. To improve the health of young children, breast-feeding was encouraged to prevent infection from unpasteurized milk. These measures improved the health and sanitation conditions of the city.

This work led to the creation of the Child Hygiene Bureau of the Depatment of Health in 1908. Baker gave up her position as assistant commissioner of health to become the director of this agency. This was the first city-funded agency of its kind.

IMPLICATION OF THE SCIENCE

Many diseases are spread by food contaminated with bacteria. One role of the Food and Drug Administration (FDA), as well as state and local health departments, is to prepare guidelines that must be followed by food-processing and meat-packing plants before their products can be sold. Food inspectors are sent to the plants to make sure the guidelines are followed. In addition, state and local health agencies inspect restaurants to make sure food is stored, cooked, and handled properly and to make sure equipment used to store, cook, and serve food is clean.

Baker brought the idea of health education into the schools with the assignment of school doctors and nurses. Under her leadership, New York City's infant death rate became the lowest of any major city in the United States.

Baker continued to receive national attention for her efforts to improve children's health. In 1916, in exchange for admission into the New York University, Bellevue Hospital Medical School, she agreed to become a lecturer. She received a Doctorate of Public Health the next year. She was the first woman to receive this degree.

For her achievements, Baker received many honors and awards. Among them was an honorary Doctor of Medical Sciences degree by the Women's Medical College of Pennsylvania.

Baker left the Child Hygiene Bureau in 1923. She remained as an advisor to the National Children's Bureau, the United States Public Health Service, and the New York State Department of Health. Baker wrote three books and more than 250 articles about child health. Most of these were published during the 1920s. She also served as the American delegate for the health committee of the League of Nations. This organization was established in 1920. One of its many goals was the improvement of world health conditions. The organization was officially dissolved in 1946 when the United Nations took its place. In 1945, Baker died of cancer at age seventy-one.

It's **YOUR TURN**

Sara Josephine
BAKER

Hands-On Activity

TRACKING THE SOURCE OF A DISEASE

MATERIALS (per group of five):

At Restaurant Stations: plastic plates, moistened filter paper or paper towel squares, forceps
At Work Stations: five paper plates, container of blue litmus paper

SAFETY

Wear safety goggles and an apron through-out the activity. Do not touch the moistened paper squares with your

PROCEDURE

1. Gather in groups of five at a work station. Have one member of the group collect materials for the group.
2. Each group member should select three numbers between 1 and 6. Record the numbers you select.
3. One member at a time, take your paper plate to the restaurant stations having the numbers you selected. At each station, use the forceps to place one piece of the filter paper on your plate. The filter paper represents food from the restaurant. Return to your work station with your plate.
4. Repeat step 3 until all members of your group have visited the restaurants.
5. Touch one piece of litmus paper to one of the paper squares. Observe the litmus paper and record your observations. Repeat this step for the other two squares on your plate.
6. If your litmus paper did not change color, the food you "ate" was safe. If your litmus paper changed color, the food you got from the restaurant was infected by bacteria. Observe whether any of the food you got was infected and record your observations. Compare your data with other members of your group.

7. Discuss with your group how to figure out which restaurant prepared the infected food. Carry out the plan your group decides on. Record the number(s) of the restaurant(s) you think is responsible for the bad food.
8. Compare your methods and results in step 7 with those of other groups.

ANALYZE AND CONCLUDE

On separate paper, write your answers.
1. How many group members got infected food?
2. Describe the plan used by your group to find out which restaurant(s) made the infected food.
3. Which restaurant(s) (by number) served infected food?
4. How did the results of your group compare with the results obtained by other groups in the class?
5. How do you think the methods your group used to identify the restaurant(s) that served the infected food are similar to those used by health officials?

34

think WORK ACT

CRITICAL THINKING Answer the following questions in complete sentences.

1. What lasting impact has the work of Dr. Sara Josephine Baker had on society?

   _____

   _____

2. Mary Mallon was a *carrier* of typhoid fever. How do you think a person who is a carrier of a disease is different from a person who is infected with a disease?

   _____

   _____

3. When Mary Mallon was identified as the person responsible for the typhoid fever outbreak in New York City, she was placed in an isolation unit. How would she be treated today?

   _____

GOING FURTHER Complete three of the following.

USING COMMUNITY RESOURCES

Write to your state Department of Health to find out what things it does as part of its health inspection of restaurants. Include your letter and the response you get in your portfolio.

RESEARCH AND REPORT

Use library resources to find out about leprosy. Write a report that describes the disease and explains how the way people who had leprosy were kept from spreading the disease was similar to the way Typhoid Mary was kept from spreading typhoid fever.

JOURNAL WRITING

Imagine that you are Typhoid Mary and you have been placed in isolation. Describe how you would feel about your situation.

COOPERATIVE LEARNING

With your group, do research to find methods used to prevent food spoilage. Also find out what regulations food service workers must follow to avoid spreading disease. Make a chart that lists each method or regulation you learn about and explain how each protects people from illness.

PERFECT YOUR SKILL

Bacteria are often classified in three groups, *cocci, bacilli,* or *spirilla,* based on their shapes. Use a dictionary to find the shapes of these types of bacteria. Make a labeled sketch of bacteria having each shape.

WOMEN OF MEDICINE

MYRA ADELE LOGAN
1908-1977

RITA LEVI-MONTALCINI
1909-

JANE WRIGHT
1919-

**SUSAN LaFLESCH
PICOTTE
1865-1915**

**GERTY CORI
1896-1957**

**JOYCELYN ELDERS
1933-**

**GERTRUDE ELION
1918-**

JANE WRIGHT

1919-

Have you ever heard the saying, "The acorn doesn't fall far from the tree?" This saying is often used to point out similarities among family members. For example, it may describe similar habits or professions of parent and child. In the case of the Wright family, this saying is certainly appropriate.

Jane Wright was born in New York City in 1919. She was the oldest of two daughters of Dr. Louis Wright and Corinne Cooke Wright. Jane's mother was a public school teacher. Her father was a surgeon and civil rights leader. He was also the director of the Harlem Hospital Cancer Research Foundation.

Like her father, Jane became a physician and cancer specialist. Jane's sister, Dr. Barbara P. Wright, is a specialist in industrial medicine. At least two other family members also held medical degrees. One, Dr. William Penn, was the first African American to earn a medical degree from the Yale University Medical School.

In her youth, Jane went to private schools in New York. After graduating high school in 1938, she earned her B.A. degree at Smith College, Northampton, Massachusetts. Jane then received a scholarship to the New York

> **Vocabulary**
> Cancer is a rapid, abnormal, continuous increase in the number of cells in a part of the body.

Medical College. In 1945, she received her M.D. degree, graduating with honors.

Wright interned at Bellevue Hospital in New York. In 1948, she became chief resident at New York's Harlem Hospital. After completing her residency, Wright went into general practice. She also worked as a doctor for the New York City school system and as a visiting physician at Harlem Hospital.

Following in her father's footsteps, Wright went to work at the Harlem Hospital Cancer Research Foundation in 1949. There, she began cancer chemotherapy research. In this research, she studied the effects of drugs on tumors. She published the results of much of her research in articles that were read by others working in this field.

After her father's death in 1952, Wright became director of the Harlem Hospital Cancer Research Foundation. She remained in this position until 1955, when she joined the faculty of the New York University Medical Center. In 1961, Wright became associate professor of research surgery.

In 1967 Wright became associate dean of the New York Medical College. She is the first African American woman physician to hold such a position. In addition to administering the medical school, Wright directed the cancer research laboratory. She also became a staff member and consulting physician at several major hospitals.

Throughout her career, Wright wrote more than 100 articles about cancer chemotherapy. Many of these articles described the methods and findings of her research. Wright's work keeps her in great demand. In 1964, she was named to the President's Commission on Heart Disease, Can-cer, and Stroke. Wright has chaired the New York Cancer Society and served on the board of directors of the New York division of the American Cancer Society and the editorial board of the *Journal of the National Medical Association*. She is also the recipient of many honors and awards. Among them is an honorary Doctor of Medical Sciences degree from the Women's Medical College of Pennsylvania. Today, Dr. Wright and her staff are still experimenting with anti-cancer drugs. She remains hopeful that a chemical cure for cancer will be found.

COOPERATIVE
COLLABORATIVE SCIENCE

Together, the different types of cancer are a leading cause of death. However, thanks to the work of researchers such as Dr. Jane Wright, more people survive cancer today than at any time in the past.

The successful treatment of cancer involves the cooperative efforts of a team of doctors who work in many specialized areas. At the head of this team is the oncologist, or cancer specialist. When cancer is suspected, the oncologist may call upon a radiologist, surgeon, and histologist to help make a diagnosis and develop a treatment plan. A radiologist is a doctor who interprets X-rays and the results of other tests such as CAT scans, PET scans, and MRIs. A surgeon may be asked to obtain a biopsy, a section of tissue that can be examined for disease. The tissue is then sent to a histologist, a specialist in preparing tissues for microscopic examination.

If cancer is found, it may be removed through surgery. Following surgery, the diseased area is often treated with radiation, a process that kills the cells of organisms. Radiation treatments are carried out by technicians working in nuclear medicine. Treatment may also involve chemotherapy—treatment with drugs. The selection of which drugs are most effective in treating a specific type of cancer is based on the work of cancer researchers such as Dr. Jane Wright.

It's YOUR TURN

Jane **WRIGHT**

Hands-On Activity

CANCER VS. NONCANCEROUS GROWTH OF CELLS

MATERIALS
pencil

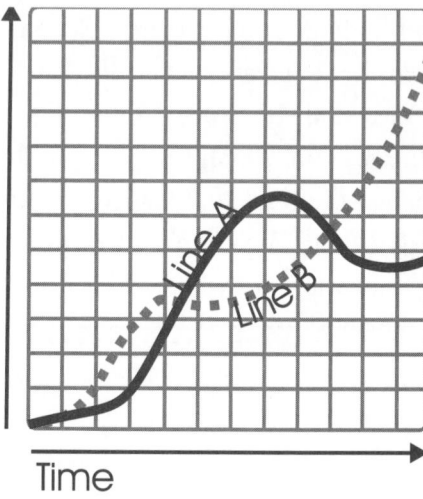

Cell Number

Line A

Line B

Time

BACKGROUND INFORMATION

Simply defined, cancer is a rapid, abnormal, continuous increase in the number of cells in a part of the body. As cancer cells become more numerous, they begin to rob healthy cells of required nutrients. Under normal circumstances, cells divide to form a pre-coded number of cells that make up a given tissue or organ. Each of these conditions can be illustrated in simple form.

PROCEDURE

1. Examine the graph. The curve of each line on the graph represents cell growth over a period of time.
2. Using the information in the Background section, identify which line (curve) represents normal cell growth and which represents cancer cell growth. Label the lines accordingly.

ANALYZE AND CONCLUDE

On separate paper, write your answers.

1. Which line did you select to represent normal cell growth? Explain why you selected this line.
2. Which line did you select to represent cancer cell growth? Explain why you labeled this line as you did.
3. How are the two lines on the graph similar? What do these similarities represent?
4. How are the two lines on the graph different? What do these differences represent?
5. What does the flattening near the end of line A indicate?
6. What does the shape of the end of line B indicate?

Name _____ Date _____

think WORK ACT

Jane
WRIGHT

CRITICAL THINKING Answer the following questions in complete sentences.

1. The prefix chemo- means "chemical." Based on this information, what do you think chemotherapy involves?

   _____

   _____

2. Wright began her medical career working in general practice. She later became a cancer researcher. How do the concerns of these two areas of medicine differ? How are they the same?

   _____

   _____

3. In what ways did the career of Jane Wright parallel the career of her father?

   _____

   _____

GOING FURTHER Complete three of the following.

CONCEPT MAPPING

Find out what methods are commonly used to treat breast cancer. Develop a flowchart that explains when these methods are used.

RESEARCH AND REPORT

Some scientists believe that many forms of cancer are caused by viruses. Research and report on these causes. What other agents have been identified as causes of cancer?

COOPERATIVE LEARNING

Find information about ways people can help reduce their risk of developing different types of cancer. Combine the information gathered by each group member to form a book.

JOURNAL WRITING

Some cancerous organs may be removed and replaced with donor organs. Such surgery is called *transplant surgery*. In your journal, explain how you think individuals should be selected for transplant surgery.

Pioneers are the first to do something noteworthy. Throughout history, there have been many pioneers. The most familiar are those who helped settle the West. More recently, pioneers have emerged in sports, business, space travel, and medicine. Few pioneers make their mark in more than one way. This was not the case with Dr. Myra Adele Logan. She accomplished many firsts during her brilliant medical career.

Myra Adele Logan was born in 1908 in Tuskegee, Alabama. She was the eighth child of Warren and Adella Hunt Logan. Warren Logan was trustee and treasurer of the Tuskegee Institute. He was appointed to these positions by the institute's founder, Booker T. Washington.

Myra completed her elementary education at Children's House, a laboratory school associated with Tuskegee Institute. She then entered Tuskegee High School, where she graduated with honors in 1923. In 1927, Myra received her bachelor's degree from Atlanta University, graduating first in her class. She then moved to New York, where she earned a master's degree in psychology from Columbia University.

Inspired by her sister Ruth, a health care provider, and her brother Arthur, a surgeon, Myra decided to study medicine. In 1929, she received a scholarship to attend the New York Medical College. She was the first African American to win such a scholarship.

Myra Adele Logan received her medical degree in 1933. She did her residency in surgery at the Harlem Hospital in New York. She worked under noted surgeon and cancer researcher Dr. Louis T. Wright.

Logan spent most of her professional life at Harlem Hospital. She developed her specialty in children's heart surgery and became known as a skilled and dedicated surgeon. In the 1940s, Logan became the first woman to perform heart surgery. In 1951, she was elected a Fellow of the American College of Surgeons. She was the first African American woman to become a member of this group.

Logan was a pioneer in the practice of group medicine. In group medicine, physicians from several specialties practice together. Logan was quick to recognize the advantages of this type of practice. Not surprisingly, she was one of the original members as well as the treasurer of the Upper Manhattan Medical Group of the Health Insurance Plan.

Logan became interested in breast cancer. In the 1960s, she began a study to detect breast tumors. She was looking for a way to find tumors that could not be detected through

> ## Vocabulary
>
> A **tumor** is an abnormal growth of tissue.

physical examination or by normal X-rays. Using a larger and slower X-ray tube, Logan was able to identify the dense breast tissue present in tumors. This process is similar to present-day mammography. She credited the use of this X-ray method with saving the lives of many women whose tumors might otherwise have gone undetected.

Throughout her career, Logan was active in many organizations. She was a member of the Physical Disabilities Program of the New York State Workman's Compensation Board. She also worked with the Planned Parenthood Association, the New York Cancer Committee, and the National Medical Association Committee of the National Association for the Advancement of Colored People.

BUILDING ON THE PAST

A great advance in the treatment of infectious diseases was the development of modern chemotherapy. Chemotherapy is the treatment of an infectious disease with a substance that inhibits the growth of, or destroys disease-causing organisms within the body. Modern chemotherapy began in the early 1900s with the research of German chemist Paul Ehrlich. In 1910, Ehrlich discovered a substance known as Salvarsan, or 606. Salvarsan, also called the Magic Bullet, effectively destroyed the organism that caused syphilis. Since Ehrlich's discovery, scientists from around the world have introduced many new antibiotics. Alexander Fleming is among the most noteworthy for his discovery of penicillin in 1928. More than thirteen years passed before this drug was completely evaluated and its use on human patients was permitted. The discovery of penicillin was followed by the discoveries of the antibiotics tyrothrin in 1939 and streptomycin in 1944.

It's YOUR TURN

Myra Adele
LOGAN

PULSE RATE BEFORE AND AFTER EXERCISE

MATERIALS (per three students) stop watch or watch with second hand, note pad, graph paper, colored pencils

BACKGROUND INFORMATION

During a physical examination, physicians check certain body functions to see how they compare to what is considered to be the norm. Included among these is the pulse rate, which is equal to the heartbeat rate. Depending upon age, gender, and state of health, this rate averages about seventy-two beats per minute.

PROCEDURE

1. Have a classmate sit on a chair in front of you. Practice locating your classmate's pulse by placing your middle and forefinger on the inside of the wrist just below the thumb.

2. For fifteen seconds, count the number of pulses you feel in your classmate's wrist. Multiply this number by four to determine the pulse rate per minute.

3. Repeat step 2 two more times. Add the results you get in each trial. Divide these results by three to get an average. Record this number.

4. Have your classmate vigorously run in place for one minute. At the end of one minute, have your classmate again sit in a chair facing you. Again determine your classmate's pulse rate.

5. Wait one minute. Again determine the pulse rate of your classmate. Repeat this step until your classmate's pulse rate returns to the value you obtained in step 3.

6. Switch roles with your classmate and repeat the entire activity. Compare the results you obtain with those of your classmate.

7. Make a graph combining the data you gathered for all members of your team.

ANALYZE AND CONCLUDE

On separate paper, write your answers.

1. How did your pulse rate change after rigorous exercise?

2. What was your pulse rate two minutes after stopping the exercise? After four minutes? After six minutes?

3. What does this number pattern suggest?

4. How long did it take following exercise for your pulse rate to return to normal?

5. The period of time it takes for your pulse to return to normal is called the *recovery rate*. What kind of information do you think recovery rate provides to a physician?

44

think WORK ACT

CRITICAL THINKING Answer the following questions in complete sentences.

1. Broad-spectrum insecticides are chemicals intended to kill a wide range of insect pests. Using this knowledge, what do you think is the role of broad-spectrum antibiotics?

_____

_____

_____

2. Identify several ways that Myra Adele Logan was a pioneer.

_____

_____

GOING FURTHER Complete three of the following.

BUILD YOUR PORTFOLIO

Use a telephone directory to determine if there are any group medical facilities in your community. What kind of services are offered by the group? How do the services provided by a medical group differ from those of a general practitioner?

RESEARCH AND REPORT

Research the work of Paul Ehrlich and the "magic bullet." Write a report that explains how his work is related to the idea of chemotherapy.

CONCEPT MAPPING

Interview a nurse or health care provider to find out what kinds of tests are performed as part of a complete blood count (CBC). Diagram your results in a concept map.

COOPERATIVE LEARNING

Collect labels from several over-the-counter medications. Review the warnings included on the labels as a group. Make a chart that summarizes the types of warnings found on the labels.

JOURNAL WRITING

Reread the biography on Myra Adele Logan. Also read the biography on Jane Cooke Wright. In your journal, describe ways in which the career paths of these two women parallel each other. How do their career interests differ?

RITA LEVI-MONTALCINI
1909-

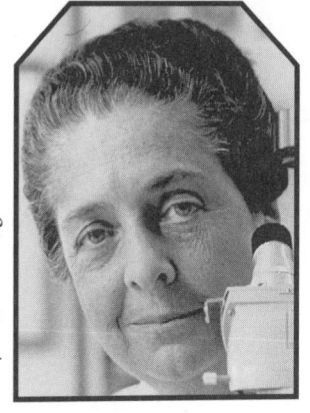

Have you ever cracked open an egg and seen a spot of blood inside? The blood spot is the beginning of the formation of the circulatory system of the developing chick. As a chick embryo develops, its heart and blood vessels form from the blood spot. If the shell remains unbroken, a live chick cracks through the shell after twenty-one days. The study of how an embryo develops from a single cell into an organism is called embryology. Rita Levi-Montalcini studies the nerve cells in developing embryos.

Rita Levi was born in Turin, northern Italy in 1909. She never married, but in true Latin tradition, Montalcini elected to take on her mother's maiden name. When she and her twin sister Paola finished the fourth grade, they were sent to finishing school. Finishing schools prepared girls to become wives and mothers. But Rita wanted to complete her education. She wanted to become a physician and had no plans of marrying.

At age twenty, Rita convinced her father to hire tutors to prepare her for the medical school entrance exam. In less than one year, Rita had mastered enough mathematics, science, Latin, and Greek to pass the exam. In 1930, she entered the University of Turin's medical school. She graduated in 1936, but stayed on for two more years to study the development of nerve cells. Then, as World War II progressed, Levi-Montalcini was forced to leave the school and was forbidden to practice medicine because of her Jewish heritage. Fearing for her safety, Levi-Montalcini went into hiding and secretly continued her research on nerve growth in chick embryos. She built a small laboratory in her home and made small operating instruments to work with nerve cells. She also developed staining techniques, methods of coloring nerve cells with dyes, to observe otherwise invisible cell parts. She published her findings in the science journals of other countries.

> ### Vocabulary
> NGF (nerve growth factor) is a protein that controls the growth and maturing process of nerve cells.

After the war, Levi-Montalcini returned to the University of Turin. Soon after, Viktor Hamburger of Washington University in St. Louis read about her work. He had conducted similar experiments. He invited Levi-Montalcini to come to the United States to discuss her research and to demonstrate her staining techniques.

Levi-Montalcini arrived in St. Louis in 1947. Hamburger asked her to join him in the research of nerve growth factor (NGF). She remained at the Washington University for more than twenty-six years. She also became an American citizen.

In 1953, Levi-Montalcini teamed up with biochemist Stanley Cohen. In 1958, she and Cohen identified NGF in the tumors of mice.

NEUROEMBRYOLOGIST

They also found NGF in the salivary glands of male mice and in snake venom. They spent the next six years gathering enough NGF for chemical analysis.

The discovery of NGF is important in cancer research because its presence may trigger the growth of cancer cells and may indicate the presence of such cells. NGF is also important in burn therapy and in the study of Alzheimer's and Parkinson's disease. In these cases, NGF may hold the key to replacing damaged nerve cells.

For their work on NGF, Levi-Montalcini and Stanley Cohen shared a Nobel Prize in physiology or medicine in 1968. That same year, Levi-Montalcini was elected to the National Academy of Sciences. Other organizations in which Levi-Montalcini is a member include the American Association for the Advancement of Science and the Society for Developmental Biology. In addition, Levi-Montalcini is the only woman who has been elected to the Papal Academy of Rome.

APPLICATION OF THE SCIENCE

Rita Levi-Montalcini developed a staining technique for use in the study of nerve cells. Stains are dyes that stick to or chemically react with certain cell parts. The development of stains and staining techniques during the mid-nineteenth century is responsible for many advances in bacteriology, histology, and cytology.

Stains are usually either acidic or basic. In general, basic stains dye cell structures that are acidic. A common basic stain is hematoxylin. Hematoxylin is often used to stain the nuclear or chromosomal material of a cell. In contrast, acid stains react with cell structures that are basic. Eosin is an acid stain that readily stains structures in the cell cytoplasm. An important feature of stains is that they can be used in combination with each other. Thus, hematoxylin and eosin are used at the same time to stain chromosomal structures and cytoplasmic material.

It's YOUR TURN

Rita LEVI-MONTALCINI

Hands-On Activity

OBSERVING THE EFFECTS OF STAINS

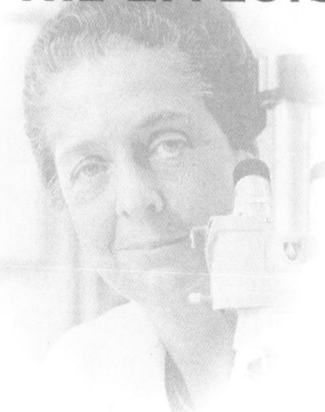

MATERIALS (per pair of students)

onion slice, forceps, microscope slide, cover slip, Lugol's iodine solution, distilled water, compound microscope

SAFETY Wear a laboratory apron throughout this activity. War safety goggles while preparing your slides.

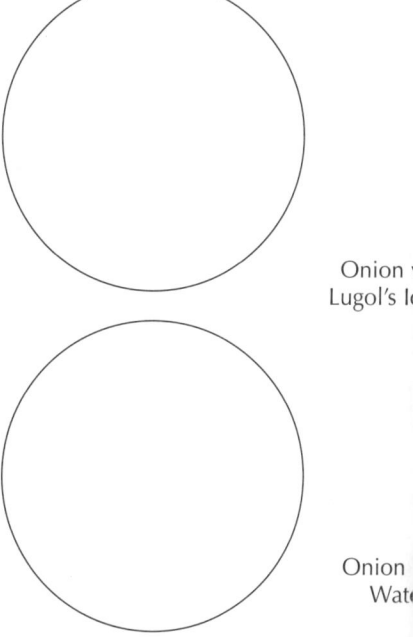

Onion w
Lugol's Io

Onion
Wate

BACKGROUND INFORMATION

The size and lack of color of cell structures makes cell parts invisible to the eye, even when using a compound microscope. One reason these structures are difficult to see is because they refract, or bend light, especially when placed in water or on glass. To make cell structures easier to recognize, scientists often make use of biological stains. Stains are chemicals that either stick to cell structures or react with the cell molecules.

PROCEDURE

1. Place a drop of Lugol's solution at the center of a clean microscope slide.
2. Using your forceps, gently pry a very thin section of onion from the onion slice. Carefully place the onion section in the iodine solution.
3. Gently place a cover slip over the onion section.
4. Examine your slide under low power with your microscope. Make a drawing of what you see in the space provided.
5. Repeat steps 1-4 using a drop of water instead of the iodine.

ANALYZE AND CONCLUDE

On separate paper, write your answers.

1. How did your observations of the onion slices placed in the two types of solutions compare?
2. How did the presence of iodine assist in your viewing of the onion skin?
3. What is the role of stains in the microscopic examination of cells or tissues?

Multicultural Women of Science ◆ Rita Levi-Montalcini   © 1996 The Peoples Publishing Group, Inc.

think WORK ACT

CRITICAL THINKING Answer the following questions in complete sentences.

1. Why did Levi-Montalcini publish the results of her work in the science journals of countries other than her own?

   _____

   _____

2. The prefix *neuro-* refers to the nervous system. Use this information to explain why Rita Levi-Montalcini is described as a neuroembryologist.

   _____

   _____

3. How did Levi-Montalcini s early work prepare her for researching NGF?

   _____

   _____

GOING FURTHER Complete three of the following.

BUILD YOUR PORTFOLIO

Redraw the cells you observed in the activity on a clean sheet of paper. Use a biology textbook or other source to label the cell parts shown in your drawings.

RESEARCH AND REPORT

Research Alzheimer's or Parkinson's disease. Find out the symptoms and suspected causes of the disease. Write a report about the disease that includes a paragraph which explains how the work of Rita Levi-Montalcini may be helpful in understanding the disease.

JOURNAL WRITING

Many people oppose the use of animals, including chick embryos, in medical research. In your journal, explain why you think animals should or should not be used in such research.

ALTERNATIVE ASSESSMENT

From your teacher, obtain two or more prepared slides of onion skin, stained with different stains. Observe the slides and make colored sketches of what you see. Use a biology textbook to identify as many cell structures as possible. Describe any similarities and differences you observe. How might you explain these differences?

GERTY CORI
1896-1957

Before a marathon, runners often eat a large meal of pasta. The runners know that pasta—a starch—can supply the body with energy. The body stores the starch until it is needed. Then the starch is changed to sugar that releases energy when it is broken down. In the 1930s, the research of Gerty and Carl Cori led to an understanding of this process.

Gerty Radnitz was born in Prague, Czechoslovakia, in 1896. Until age ten, she was schooled at home. Then she went to a private school. By the time she was sixteen, Gerty knew she wanted to study medicine; however, at that time, the schools for young women did not teach many of the subjects needed for medical school. For two years, Gerty studied on her own to learn the Latin, math, physics, and chemistry needed for entry into medical school. Her hard work paid off and she entered the German Medical School of the University of Prague at age eighteen.

While in medical school, Gerty decided to devote her life to research. She met another student, Carl Ferdinand Cori, who shared her interest in research. In 1920, Gerty received her medical degree and married Carl Cori.

The Coris found that there were few research jobs in Europe. In 1922, Carl emigrated to the United States to begin work as a biochemist. His job was with the New York State Institute for the Study of Malignant Diseases in Buffalo. (Today, the Institute is known as the Roswell Institute.) Gerty emigrated six months later when she got a job as an assistant pathologist at the same institute. Six years later, Gerty became an assistant biochemist at the institute. About the same time, both Coris became citizens of the United States.

The Coris grew interested in carbohydrate metabolism. Their research at Buffalo had little to do with this topic, so they decided to leave to pursue their interest. In 1931, Carl became chairman of the pharmacology department of the Washington University School of Medicine. Gerty obtained a position as an assistant at the same university. Eventually, both Gerty and Carl moved to the biochemistry department, but Gerty did not become a full professor in the biochemistry department until 1947. That same year, the Coris shared the Nobel Prize in physiology or medicine with Bernardo Houssay of Argentina.

The Coris won the Nobel Prize for their discovery of the enzyme that changes glycogen, an animal starch, into glucose, a sugar. They also discovered how this reaction and the reverse

> ### Vocabulary
> Biochemistry is the branch of biology that studies substances that make up or are used by living things.

reaction take place. Houssay shared in the prize for showing how the pituitary gland is involved in sugar metabolism.

Gerty Cori was the first American woman to win a Nobel Prize. As a team, the Coris were the third married couple to share the award. Previous couples who shared the award were Pierre and Marie Curie and Irene and Frederic Joliot-Curie.

The same year the Coris won their Nobel Prize, Gerty learned she had a rare blood disease. Her bone marrow did not make enough red blood cells. For the next ten years, she needed frequent blood transfusions; however, in spite of her illness, she continued to work. During this time, she identified many new diseases related to glycogen storage. She also determined the molecular structure of glycogen. This work aided researchers in their studies of diabetes.

APPLICATION OF THE SCIENCE

If you were asked to prepare a balanced menu, how would you go about this task? In a balanced diet, you need certain amounts of fats, proteins, carbohydrates, vitamins, and minerals. These substances, called nutrients, are present in the foods you eat.

Scientists have devised tests to identify the presence and amounts of each of these nutrients in various types of foods. For example, iodine is used to test for the presence of starch in food. If iodine is added to a food containing starch, the iodine changes from an amber to a blue-black color. Biuret solution is a substance used to test foods for proteins. While some chemical tests identify the presence of a nutrient, other more sophisticated tests can identify the amount of the nutrient present in different types of food. The results of such nutrient tests are used by dieticians and nutritionists to help them prepare balanced menus.

It's YOUR TURN

Hands-On Activity

CHANGING STARCH TO SUGAR

MATERIALS (per group of four)

starch mixture, three test tubes, test tube holder, test tube rack, hot plate, beaker, Benedict's solution, bacterial amylase, dropper, glass marking pencil, water, clock or watch, oven mitt or pot holder

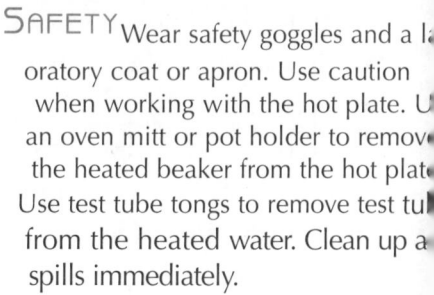

SAFETY
Wear safety goggles and a laboratory coat or apron. Use caution when working with the hot plate. Use an oven mitt or pot holder to remove the heated beaker from the hot plate. Use test tube tongs to remove test tubes from the heated water. Clean up all spills immediately.

BACKGROUND INFORMATION

Digestion is a life process through which the nutrients in the foods you eat are broken down into a form that can be used by your body. Enzymes are chemicals formed by the body that help speed this reaction. Starch is a complex nutrient that is acted upon by an enzyme called *amylase*, which is produced by the pancreas. When released in the body, amylase helps change starch into sugar that can be used by the body.

Table 1—Test Results

Test tube	A	B	C
Color of liquid			

PROCEDURE

1. Use the glass marking pencil to label the three test tubes *A*, *B*, and *C*. Set the test tubes in the test tube rack.
2. Half-fill test tubes A and B with the starch mixture. Half-fill test tube C with water.
3. Add three-four drops of bacterial amylase to test tubes B and C.
4. Add some water to the beaker and heat the beaker on the hot plate.
5. One at a time, add three-four drops of Benedict's solution to each test tube. Place the test tubes in the water bath to heat them for one minute. If the mixture in the test tube begins to boil, remove it from the heat immediately.
6. Observe the color of each mixture after it is heated. Record the color in the table.

ANALYZE AND CONCLUDE

On separate paper, write your answers.
1. Benedict's solution turns from blue to orange when sugar is present. Which test tubes contained sugar?
2. Where did the sugar that was present in the test tube after it was heated come from?
3. People who have diabetes must limit the amount of sugar they take into their bodies. Using your observations from this activity, explain why a person with diabetes should also control the amount of starch he or she takes into his or her body.

think WORK ACT

CRITICAL THINKING   Answer the following questions in complete sentences.

1. Insulin is a hormone that controls the breakdown and use of sugar (glucose) in the body. People with diabetes do not produce enough of this hormone. How was the work of the Coris related to the study of diabetes?

_____

2. Bernardo Houssay won his Nobel Prize for showing how the pituitary gland was involved with sugar metabolism. How was the work of Houssay related to the work done by the Coris?

_____

3. Many fruits, such as bananas, contain starch. Such fruits get sweeter as they ripen. Describe the process that must be taking place in the fruit as it ripens.

_____

_____

GOING FURTHER   Complete three of the following.

BUILD YOUR PORTFOLIO

Collect fifteen to twenty food labels. Analyze the labels to determine which nutrients are most commonly found in packaged foods. Write a brief summary of your findings.

JOURNAL WRITING

Diabetes is a life-long illness that requires daily medication and strict attention to diet and exercise. In your journal, describe how having a disease such as diabetes would change your life. Additional research may be necessary.

RESEARCH AND REPORT

Do research on the life of Bernardo Houssay. Use this information to write a biographical sketch of Houssay's life and Nobel Prize-winning work.

COOPERATIVE LEARNING

Have each member of your group bring in samples of three foods. Test each food for fat by touching the food to a piece of brown paper. A brown paper bag will do. The formation of a translucent (oily looking) spot on the paper indicates the presence of fat. Organize your findings in a table.

ALTERNATIVE ASSESSMENT

Iodine can be used to indicate the presence of starch in food. If starch is present, iodine reacts with the starch and becomes blue-black in color. Choose 15 different foods to test for starch. Test each food by placing one drop of iodine on the food sample and observing the color of the drop. Record your findings in a table. Wear safety goggles, a laboratory coat or apron, and protective gloves for this test.

Not long ago, kidney disease, some forms of leukemia, and some diseases caused by the herpes virus were fatal. Today, there is hope for recovery from each of these problems. Many people with these illnesses look forward to rewarding lives due to the creative mind of biochemist Gertrude Belle Elion.

Gertrude Elion was born in New York City in 1918. At age fifteen, she saw her grandfather die of cancer. Elion's grief prompted her to devote her life to medical research.

Elion attended public school in the Bronx. She then went to Hunter College where she majored in chemistry. In 1937, she graduated college with high honors; however, she lacked the money needed to continue her education. To raise the money she needed, Elion took a variety of short-term jobs. One of these jobs was as a substitute high school science teacher. She also worked in various food laboratories.

While she worked, Elion attended night school at New York University. In 1941, she received a master's degree in chemistry. Around the same time, her fiancé died. Elion decided never to marry and to devote her life to her work.

In 1944, Elion was hired by Burroughs Welcome Company, a New York pharmaceutical firm. She worked as a research assistant for George Hitchings, a scientist who was trying to find drugs to treat cancer.

When Elion and Hitchings first began their work together, they tested drugs using a trial-and-error approach. A drug was tested to see if it worked against a certain disease. If the drug did not work, another drug was tried. Although this approach met with some success, Hitchings decided to try a new approach. The new method focused on how different drugs acted on individual cells rather than their effectiveness in curing a disease.

Building on Hitchings' methods, Elion developed ways to change the molecular structure of the nucleic acids in disease-causing cells without affecting the molecular structure of nucleic acids in healthy cells.

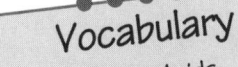

Vocabulary

Nucleic Acids
give cells the information they need to produce chemicals that are vital to the human body.

In 1950, Elion synthesized two small molecules that are part of nucleic acids. Compounds formed using these molecules proved effective in treating leukemia. An altered form of the drug was used with organ transplant patients. This drug suppressed the immune system, preventing the body from rejecting the transplanted organs. The same compound could also be used to treat anemia, hepatitis, and severe rheumatoid arthritis. During her research career, Elion developed numerous compounds to treat other diseases including gout and herpes.

BIOCHEMIST

Elion retired from the Burroughs Welcome Company in 1983. Since her retirement, other drug companies have used the techniques developed by Elion and Hitchings to create new drugs. One such drug is AZT, the first drug developed for the treatment of acquired immune deficiency syndrome (AIDS). AZT is used successfully with many AIDS patients.

Elion has received numerous awards and honors, including induction as the first woman into the National Inventor's Hall of Fame, the National Women's Hall of Fame, and the Engineering and Science Hall of Fame. Elion and George Hitchings shared the Nobel Prize in physiology or medicine with a British biochemist in 1988. Elion was seventy at the time.

PERSPECTIVES

Throughout history, diseases such as the Bubonic plague, poliomyelitis, tuberculosis, and more recently AIDS, have had devastating impacts on society. Scientists worldwide work hard to develop drugs to prevent and treat such diseases. Many of these drugs come into use each year; however, only a few benefit society enough to earn recognition for the scientists who develop them. The timeline indicates when some of these drugs were developed.

TIMELINE

1796   Dr. Edward Jenner develops a vaccine for small-pox.

1892   Use of the diphtheria antitoxin developed by Emil Adolf von Behring and Shibasaburo Kitasato begins.

1910   Paul Ehrlich develops salvarsan, a drug used to treat syphilis.

1928   Alexander Fleming accidentally discovers the Penicillium mold.

1939   Ernst Chain and Howard Florey produce pure penicillin from the mold discovered by Fleming.

1948   Elizabeth Hazen and Rachel Brown develop Nystatin, the first fungicide safe for use in humans.

1953   Jonas Salk begins testing a vaccine for poliomyelitis.

1957   Dr. Albert Sabin produces an oral polio vaccine.

1965   Drs. Paul Parkman and Harry M. Meyer develop a vaccine for rubella (German measles).

1987   AZT, a drug used to treat AIDS, is made using the pioneering research methods developed by Gertrude Elion.

Gertrude
ELION

It's YOUR TURN

Hands-On Activity

APPROACHES TO PROBLEM SOLVING

MATERIALS
(per individual)
key-operated lock
assortment of keys

BACKGROUND INFORMATION

There are many approaches to solving problems. One approach involves trial-and-error. In trial-and-error, different attempts are made to solve a problem until an attempt that works is discovered. Often trial-and-error approaches to solving problems do not involve reasoning.

PROCEDURE

1. Devise a plan for finding the key that opens your lock. Describe your plan in writing in your notebook.
2. Carry out the plan you described in Step 1. Observe how many keys you use before finding the key that opens the lock.

ANALYZE AND CONCLUDE

On separate paper, write your answers.

1. How did you go about finding the key that opened the lock?
2. How many keys did you try before finding the key that opened your lock?
3. Describe another approach you could use to find the key that opens the lock.
4. What is the main difference between the method you used to open the lock and the method you described in question 3?

think WORK ACT

CRITICAL THINKING Answer the following questions in complete sentences.

1. How did the methods for developing drugs used by Gertrude Elion and George Hitchings differ from those used by other pharmaceutical workers?

   _____

   _____

2. Describe the importance of the work done by Gertrude Elion.

   _____

   _____

3. Why do you think it might be important for a food company to hire chemists to test their products?

   _____

   _____

GOING FURTHER Complete three of the following.

COMMUNITY RESOURCES

Research or write to the U.S. Food and Drug Administration or the public relations department of a pharmaceutical company. Find out how a drug is approved for use in this country. Summarize the steps involved in the approval process.

PERFECT YOUR SKILL

Interview your physician, a parent, guardian, or school nurse to find out what vaccinations you have been given and at what age each vaccine was given. Find out which vaccines were required for you to be allowed to enroll in school. Organize your findings in a table.

JOURNAL WRITING

In your journal, identify the disease for which you would most like to see scientists develop a cure. Explain why you chose this disease.

COOPERATIVE LEARNING

Have each member of your group research one of the following diseases: diphtheria, measles, mumps, tetanus, and chicken pox. For each disease, find the cause of the disease, its symptoms, its treatment, whether a vaccine exists for the disease, when the vaccine was developed, and the name of the person(s) who developed it. Organize your findings in a table.

RESEARCH AND REPORT

Do library research on poliomyelitis. Find out how the development of a polio vaccine affected the number of cases of poliomyelitis that occur each year. Write a brief report of your findings.

JOYCELYN ELDERS
1933-

The position of Surgeon General of the United States is one of honor, authority, and responsibility. The job of the surgeon general is to improve awareness of health problems such as those related to smoking, alcohol, and sexually transmitted diseases. The surgeon general is also responsible for health issues concerning women and minorities and for the President's Council on Physical Fitness and Sports.

In 1990, President George Bush nominated Antonia Novello to the position of United States Surgeon General. On March 9, 1990, Dr. Novello became the first Latin American woman to serve as surgeon general. She held this post until 1993, when President Bill Clinton nominated Joycelyn Elders to this position. With her appointment, Dr. Elders became the second woman to serve as Surgeon General of the United States. She also became the first African American to hold the post.

Like many physicians, Elders specialized. Her field of specialization was pediatric endocrinology. In this field, Elders treated children who had diseases or disorders involving the endocrine, or ductless, glands. Common disorders of this sys-

Vocabulary

endocrinology is the study of the various ductless glands and their secretions

tem include juvenile diabetes and dwarfism. These diseases result from the body not producing enough of certain chemical substances called hormones.

Prior to her confirmation, Elders was director of the Arkansas Department of Health for nearly six years. She was appointed to that position by then Governor of Arkansas Bill Clinton. She has also worked as a professor of pediatrics at the University of Arkansas Medical School.

Joycelyn Elders, one of eight children, was born Minnie Joycelyn Jones in Schaal, Arkansas. Her parents were tenant farmers. At age fifteen, Joycelyn received a scholarship to Philander Smith College of Arkansas from the United Methodist Church. Elders worked as a maid to support herself while in college. She graduated in only three years. She then joined the U.S. Army as a first lieutenant. After completing her military service, Elders went to the University of Arkansas Medical School, from which she graduated in 1960.

Elders has received many honors and awards. Among them are the National Governor's Association Distinguished Service Award, the American Medical Association's Dr. Nathan Davis Award, and the National Coalition of 100 Black Women's Candace Award for Health Science. Elders also has honorary doctor of medical science degrees from Morehouse College, Yale University, and the University of Minnesota. In addition to her work as a physician and as Surgeon General, Elders has written more than 150 articles on hormone-related ill-

nesses. Many of these articles focus on hormones and their link to children's growth patterns.

PERSPECTIVES

During the 1930s, a great number of deaths caused by lung cancer prompted scientists to research the relationship between smoking and lung cancer. By the 1950s, enough data had been gathered for scientists to link smoking to several diseases, including lung cancer, chronic bronchitis, emphysema, and heart disease. In 1956, the United States Public Health Service, through its director, the Surgeon General, established a joint commission to study the effects of smoking on health. The commission concluded that there was a causal relationship between smoking and lung cancer.

In 1961, President John F. Kennedy commissioned a panel to again investigate the effects of smoking on health. Still another commission was assembled in 1962 by then Surgeon General Luther L. Terry. After an 18-month study, the commission concluded that "cigarette smoking is a health hazard of sufficient importance in the United States to warrant an appropriate and immediate action." In 1969, Surgeon General William H. Stewart issued the following statement, "We know of no organized medical or scientific body in the world which states that cigarette smoking is not a health hazard. . . . "

More recently, Surgeons General C. Everett Koop, Antonia Novello, and Joycelyn Elders have continued to focus on the dangers and health risks associated with cigarette smoking and tobacco use. Warning labels on cigarette packages express more than a possible connection to lung and heart disease. Label warnings, combined with limited or restricted advertisements, are all intended to help make the public aware of the dangers of cigarette smoking.

> SURGEON GENERAL'S WARNING:
> Cigarette Smoke
> Contains Carbon Monoxide.

It's YOUR TURN

Hands-On Activity

SMOKING AND HEART RATE

MATERIALS (per student)

note pad, pencil, watch with a second hand, graph paper

SAFETY

To avoid exposure to secondhand smoke, have the smoker record his or her own pulse rates while you are in a different room.

BACKGROUND INFORMATION

Nicotine, in its pure form, is a clear, colorless liquid. It is a powerful, quick-acting poison that is sometimes used as an insecticide or fungicide.

Nicotine is harmful to the circulatory system. It decreases the size of the openings of blood vessels. This is especially dangerous when it involves the coronary arteries   the blood vessels that provide blood to the heart. Reduction in the size of these arteries causes the heart rate to increase and raises blood pressure. Loss of blood to the heart muscle is a major cause of heart attacks.

PROCEDURE

1. At home, identify a friend or family member who is a smoker. Have the smoker take his or her pulse for one minute before they smoke a cigarette. (To obtain the pulse rate, count the number of pulses you feel in 15 seconds and multiply this number by four.) Record the pulse rate.

2. Have the smoker record his or her pulse rate after they have lit and inhaled from a cigarette at least four times.

3. After the smoker has put out the cigarette, have him or her take and record his or her pulse rate every minute until the pulse rate returns to the rate obtained in step 1.

4. In class, plot the readings provided to you by the smoker on a sheet of graph paper.

5. Take and record the pulse of a classmate. Take the pulse again each minute for 5 minutes. Plot this data on your graph.

6. Compare your results with those of your classmates.

ANALYZE AND CONCLUDE

On separate paper, write your answers.

1. How did the pulse rate of the smoker change before and immediately after smoking?

2. How much time was required for the pulse rate of the smoker to return to normal?

3. How did the pulse rate of the smoker compare with your classmate s pulse rate?

4. What effect, if any, does smoking seem to have on pulse rate?

think WORK ACT

CRITICAL THINKING Answer the following questions in complete sentences.

1. Why is the job of the Surgeon General important?

   _____

   _____

2. Many people believe the job of the Surgeon General should be eliminated to make government smaller and save tax dollars. Explain whether you think the job should be kept or eliminated.

   _____

   _____

3. What qualifications do you think a candidate for Surgeon General should have? Explain.

   _____

   _____

GOING FURTHER Complete three of the following.

BUILD YOUR PORTFOLIO

Find the names of the people who have served in the position of Surgeon General during the last 10 presidential administrations. Make a timeline of your findings that shows when each person served in the position.

CONCEPT MAPPING

Find out what processes are involved between the time a person is nominated to the position of Surgeon General and the point at which they take office. Organize your findings in a concept map.

JOURNAL WRITING

In your journal, describe a health care issue that is important to you. Explain why you selected this issue.

COOPERATIVE LEARNING

Select a public health issue that you think people should know more about. Design a campaign to make people more aware of the issue. Prepare the materials needed to carry out your campaign.

ALTERNATIVE ASSESSMENT

Using the activity on p. 60 as a model, design an experiment, or do research, to find out what effects other forms of tobacco (such as, pipes, cigars, chewing tobacco, snuff) have on the pulse rate of the person who uses these products. Create a graph that shows how the results you obtain compare to those you obtained in the activity or write a report of your findings.

SUSAN LaFLESCHE PICOTTE
1865-1915

Within some societies, certain members of the group are thought to have supernatural powers. Among these powers are the ability to bring rain and to heal those who are ill or wounded. In some cultures, the healer is called a witch doctor. Among Native Americans in the U.S., the healer is often called a shaman. Just before the turn of the century, the Omaha Indians in Nebraska had their own medical doctor. The doctor was Susan LaFlesche Picotte, a woman and a respected member of the community.

Susan LaFlesche was born on the Omaha reservation in northeastern Nebraska in 1865. She was the youngest of five children born to Chief Joseph LaFlesche (Iron Eye) and his wife Mary Gale (One Woman). Both parents were part white and part Native American. They stressed the importance of education and sent Susan and her older sister, Susette, to schools outside the reservation. Chief Joseph also respected the culture and advanced farming methods of the settlers and tried to model his village after theirs. For this, his people gave the village the nickname, "Make Believe White Man's Village."

Susan and Susette attended the Elizabeth

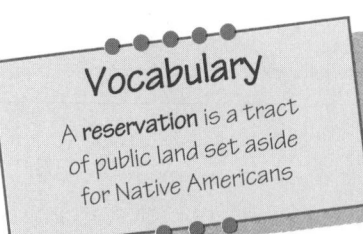

Vocabulary
A **reservation** is a tract of public land set aside for Native Americans

Institute for Young Ladies in New Jersey. Susette became an advocate for Native American rights. She traveled to Washington, speaking against the wrongs done to Native Americans. Susan was inspired by the sick and dying among her people. She chose medicine as her profession.

In 1887, Susan entered the Women's Medical College of Pennsylvania. She completed the three-year course in two years, graduating first in a class of thirty-six. Susan became the first Native American woman to receive a medical degree. She was only twenty-three years old.

Susan returned to the Omaha reservation as a physician in 1889. She provided medical services to both Native American and non-Native American children. Her patients were spread out across the reservation, but the distance and rugged terrain never kept her away. She rode on horseback, often in freezing temperatures or in storms to attend to the sick. While on the reservation, she met Henry Picotte, a Yankton Sioux. The couple was married in 1894. Henry Picotte died in 1905. Despite her grief, Susan continued her duties as a physician and taught people how to improve their health.

During her twenty-five-year service as a physician and teacher, Dr. Picotte had provided medical service to almost everyone on the reservation. She was widely loved and respected. The

PHYSICIAN

Omaha people considered her a leader during a period in history when women did not occupy positions of authority.

Besides her duties as a physician, she was the chairperson of the local Board of Health. In 1913, Dr. Picotte founded a hospital at Walthill, a town established on the Omaha reservation. She also helped to organize the County Medical Society there. Both the medical society and the hospital provided needed medical services to the people on the reservation. Following her death in 1915, the hospital was renamed in her honor.

BUILDING ON THE PAST

Early Native American populations identified the medicinal values of roots, stems, leaves, and bark of a variety of plants. Over time, many of the remedies identified by these cultures have not only been refined, but have become an important part of modern medical treatment. Aspirin, digitalis, and quinine are three common drugs that have their roots in Native American culture. Aspirin, which comes from the bark of the willow, is a common treatment for minor pain. More recently, aspirin has been used as a blood thinner in people suffering from heart problems. Digitalis is a heart medicine derived from the foxglove plant. Quinine, a substance derived from the cinchona tree, is used to treat malaria. Native American cultures also discovered the use of plants as sources of several pain killers and anesthetics. Often, these plant substances were used in both medicine and hunting.

It's YOUR TURN

Hands-On Activity

BREATHING RATE BEFORE AND AFTER EXERCISE

MATERIALS

(per three students)

stop watch or watch with second hand, note pad, graph paper, colored pencils

BACKGROUND INFORMATION

During a physical examination, physicians try to determine certain normal body functions. Included among these are breathing rate and heartbeat rate. In measuring breathing rate, the normal number of breaths per minute is about eighteen. In addition to the number of breaths per minute, a physician may also observe movements of the ribs, chest, and shoulders. These movements very often indicate the presence of such respiratory diseases as emphysema, bronchitis, and asthma.

PROCEDURE

1. Have a classmate sit on a chair in front of you. For a period of one minute, count the number of times your classmate inhales (breathes in) and exhales (breathes out). Each inhalation and exhalation is counted as one breath.

2. Repeat step 1 two more times. Add the results you get in each trial. Divide these results by three to get an average. Record this number.

3. Have your classmate vigorously run in place for one minute. At the end of one minute, have your classmate sit in a chair facing you. Take a count of the breathing rate for the first minute your classmate is seated.

4. Wait two minutes. Again determine the breathing rate of your classmate. Repeat this step until your classmate s breathing rate returns to the value you obtained in step 2.

5. Switch roles with your classmate and repeat the entire activity. Compare the results you obtain with those of your classmate.

6. Make a graph of the data you gathered for all members of your group.

ANALYZE AND CONCLUDE

On separate paper, write your answers.

1. How did your breathing rate change after rigorous exercise?

2. What was your breathing rate two minutes after stopping the exercise? After four minutes? After six minutes?

3. What does this number pattern suggest?

4. How long did it take following the exercise for your breathing rate to return to normal?

think WORK ACT

CRITICAL THINKING Answer the following questions in complete sentences.

1. Why do you think the healers of some societies are called witch doctors?

   _____

   _____

2. In what ways are the career paths chosen by Susan LaFlesche Picotte and her sister Susette LaFlesche similar?

   _____

   _____

3. What obstacles do you think made it difficult for Susan LaFlesche Picotte to provide medical care to all the Omaha Indians.

   _____

GOING FURTHER Complete three of the following.

BUILD YOUR PORTFOLIO

Do research to find the names of five plants that are useful for making medicines. For each plant, find out where it grows, what medicine is made from it, and what disease or diseases the medicine is used to treat.

RESEARCH AND REPORT

Conduct research to find out what diseases were common at the time that Susan LaFlesche Picotte was practicing medicine. How were these diseases treated at the time? How are they treated today?

COMMUNITY RESOURCES

Take the graph you made in the activity on p. 64 to a respiratory therapist or physician. Find out how your data compares with statistics for other people in your age group. Prepare a summary of your findings.

COOPERATIVE LEARNING

Imagine your group represents a local board of health. As a group, identify what you think are the medical needs of your community and what resources you would need to address these needs. Consider such factors as family planning, sanitation, pollution, and spread of infectious diseases.

JOURNAL WRITING

Imagine you lived at the same time as Susan LaFlesche Picotte. In your journal, describe some forms of technology that are available today, that were not available to Dr. Picotte. How might these forms of technology have been helpful to Dr. Picotte?

WOMEN OF

EARTH SCIENCE

**SYLVIA
EARLE MEAD
1935-**

**WINIFRED GOLDRING
1888-1972**

**FLORENCE BASCOM
1862-1945**

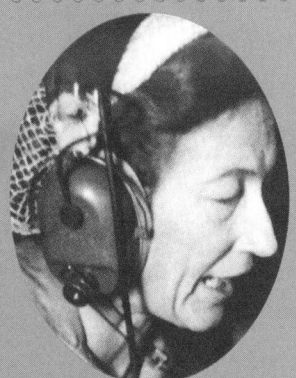

**JOANNE MALKUS
SIMPSON
1923-**

**HENRIETTA
SWAN LEAVITT
1868-1921**

**ANNIE JUMP CANNON
1863-1941**

AND

SPACE
SCIENCE

**MAE JEMISON
1956-**

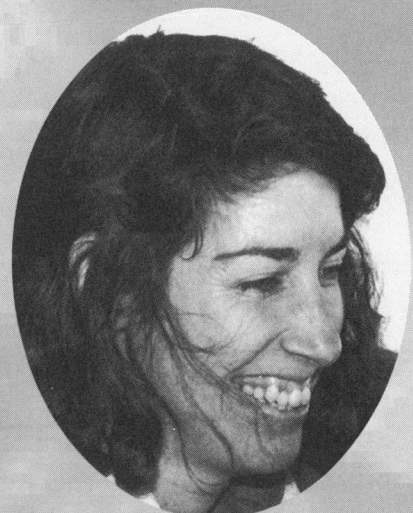

**ELLEN OCHOA
1958-**

WINIFRED GOLDRING

1888-1972

Have you ever searched for fossils? Fossils are the preserved remains or traces of living things from the past. Fossils help scientists uncover information about things that no longer exist. They also help scientists learn about how living things have changed over time. Winifred Goldring was a paleontologist who made collecting and studying fossils her career.

Winifred Goldring was one of nine children of Frederick and Mary Goldring. Winifred was born in a small town near Albany, New York, in 1888. Soon after her birth, the family moved to Slingerlands, another suburb of Albany. This area was rich in rocks, minerals, and fossils. While growing up, Winifred loved to collect rocks, minerals, and fossils. With her curiosity, it should have surprised no one that she decided to become a paleontologist.

Winifred was a bright child. In 1905, she graduated from high school at the top of her class. She then went to Wellesley College in Massachusetts. Winifred's high grades led to her acceptance into the distinguished Phi Beta Kappa honor society.

In 1909, Goldring received her bachelor's

Vocabulary

Petrology is a branch of geology that deals with the composition and formation of rocks.

degree. She remained at Wellesley, working as an assistant to geology professor Elizabeth Fisher. Goldring earned her masters degree from Wellesley in 1912. During the next two years, Goldring worked as an instructor of petrology and geology.

During the summer of 1913, Goldring took some graduate courses at Columbia University in New York. The following year, she was hired as a science resource person by the New York State Museum in Albany.

On her own, from 1916 to 1923, Goldring researched and collected Devonian plants. The Devonian period, also called the Age of Fishes, occurred from 345 to 395 million years ago. Her reasearch would serve her well.

One part of Goldring's job was to develop science exhibits for the museum. Goldring took her job very seriously. She studied the exhibits put together by other workers at the museum. Goldring found these exhibits dull and uneducational for people who were not trained in science. She developed new exhibits that were interesting and understandable to the average person. The success of Goldring's work was due largely to her earlier research experiences and to her use of paintings and creative fossil displays. Her artwork helped to show what the natural surroundings of the fossil organisms looked like when the organisms lived.

Goldring became well-known for her exhibits, and many were used as teaching tools.

Because of her success, Goldring was appointed Associate Paleontologist at the museum in 1920.

Goldring spent time writing about her field. In 1923, her book *The Devonian Crinoids of the State of New York,* was published. This book established Goldring as an expert on crinoids. Crinoids, also called sea lilies, were ancient sea animals once thought to be plants. Other books published by Goldring include her *Handbook of Paleontology for Beginners and Amateurs (Volume 1)* in 1929 and *(Volume 2)* in 1931. In 1933, she published *Guide to the Geology of John Boyd Thacher Park.* These books were used in the classroom and as study guides.

Goldring became the state paleontologist for New York in 1939. She was the first woman to hold this position, and remained in the job until her retirement in 1954. Throughout her life, Goldring received many other honors. She was the first woman president of the Paleontology Society, in 1949. The following year, she served as vice president of the Geological Society of America.

Despite her success in her field, Goldring did not encourage other women to enter the field of paleontology. Like most fields of science in her time, paleontology was considered a field for men. Women who chose to work in this field were often treated harshly and snubbed by their male counterparts. In addition, work was scarce for women paleontologists.

Goldring never married. She often claimed that she could not find anyone who was as attractive to her as her work. After her retirement in 1954, she lived in the family home with several of her sisters. She spent her leisure time reexploring the region of her youth. Two days before her eighty-third birthday in 1971, Winifred Goldring died quietly at home.

BUILDING ON THE PAST

Women paleontogolists and researchers have discovered dinosaurs and human ancestors, too.

1811 Mary Anning finds the first complete skeleton of an *Ichthyosaur,* a dolphion-like reptile.

1822 Mary Ann Mantell finds fossil parts of an Iguandon, a plant-eating reptile.

1959 Mary Leakey discovers parts of *Australopithecus bosei,* a two-million-year-old human ancestor.

1964 Mary and Louis Leakey find parts of *Homo habilis,* a 3.6-million-year-old human ancestor.

1977 T. Maryaska describes *Saichania* and *Tarchia,* bone-headed dinosaurs from Mongolia.

1986 Graciela Bachotey, Olga Gimenez, and others name the dinosaur *Xenotarsaurus;* Angela Milner names *Baryonyx;* Emily Cobabe and D.E. Fastovsky describe *Ugrosaurus.*

It's YOUR TURN

Winifred GOLDRING

MAKING FOSSIL MOLDS AND CASTS

MATERIALS (per group of three or four)

newspapers, modeling clay, shoe box lid, object to be used as fossil, small paint brush, mineral oil, plaster of Paris, water, paper cup, stirrer

SAFETY

Wear safety goggles and a laboratory coat or apron throughout this activity

BACKGROUND INFORMATION

Many different types of fossils indicate the presence of organisms. For example, the hard parts of organisms, such as bones, shells, and teeth that don t decay when an organism dies are fossils. Another type of fossil forms when the remains of an organism decay, leaving behind only an impression. Such a fossil is called a *mold*. A footprint is a mold. A mold may fill with mineral-rich water. As the water evaporates, the minerals left behind harden to form a *cast* that has the shape of the original animal or part that formed the impression.

PROCEDURE

1. Cover your work surface with newspapers. Write your group s name or number in the shoe box lid. Place the clay in the lid. Shape the clay into a 10-cm square about 2-cm thick.
2. Press an object such as your hand or a coin into the surface of the clay to make an impression. Carefully remove the object from the clay.
3. Use a brush to coat your clay impression with mineral oil.
4. In a cup, mix the plaster of Paris according to the instructions on the package or the instructions given by your teacher.
5. Pour the plaster of Paris mixture into the impression in the clay. Allow the plaster of Paris to harden, undisturbed, overnight.

6. Gently separate the plaster of Paris from the clay so as not to break the plaster or disturb the impression.

ANALYZE AND CONCLUDE

On separate paper, write your answers.
1. What type of fossil mold or cast does the plaster of Paris object represent?
2. What type of fossil mold or cast does the clay impression represent?
3. How are the two types of fossils similar? How are they different?
4. How does each fossil compare with the original object from which it was made?

think WORK ACT

CRITICAL THINKING Answer the following questions in complete sentences.

1. How do fossils teach about the past?

   _____

   _____

2. How do fossils show changes in living things?

   _____

   _____

3. How might fossils provide scientists with information about how the surface or physical structure of some areas of Earth have changed over time?

   _____

GOING FURTHER Complete three of the following.

BUILD YOUR PORTFOLIO

Find out what a crinoid looks like. Make a drawing of a crinoid for your portfolio. In a paragraph, explain why some people may describe the animal as being plantlike.

ALTERNATIVE ASSESSMENT

Use modeling clay to record footprints (bare feet) of 10-12 people of various heights and weights. Record the height and weight of each person on a sheet of paper. Make measurements of the footprints (length, width, and depth), and determine if there is any correlation between foot size, height, and weight. Write a report of your findings.

JOURNAL WRITING

In your journal, identify an organism from long ago that you find interesting. Explain why this organism interests you.

COOPERATIVE LEARNING

The geologic time scale divides Earth's history into eras, periods, and epochs. Make an illustrated geologic time scale showing these divisions and the organisms living during each time. Use library resources for your research. Each group member should research and prepare materials for one era.

RESEARCH AND REPORT

Conduct research to find out what fossils of dinosaurs or other ancient animals or plants have been found in your state or a neighboring state. Prepare a map of the state that shows where the finds were made. Create a key to identify the types of organisms found.

FLORENCE BASCOM
1862-1945

Earth is about 4.5-billion years old. Since it first formed, Earth has undergone many changes. Rivers cut deep gorges into the land, carrying away rocks and soil. Sediments dropped by rivers build up the land. At different times, seas have moved inland to cover coastal areas and ocean waves pounded the shorelines, wearing them away. Lava from volcanoes form mountains, while glaciers erode other mountains. The materials resulting from these activities have formed different kinds of rocks. The study of rocks and the events that create and destroy landforms are concerns of the science of geology. Florence Bascom made this science her life's work.

Florence Bascom was born in Williamstown, Massachusetts, in 1862. Her father, John Bascom, was a professor at Williams College. Her mother was a strong supporter of women's rights and education for women. Both parents were also interested in science, which may have influenced Florence's career choice.

In 1874, John Bascom became president of the University of Wisconsin and the family moved to Madison, where Florence attended high school. She graduated at age fifteen and enrolled at the University of Wisconsin.

> ### Vocabulary
> **Geology** is the study of the composition, structure, and history of the earth, moon, and other planets.

During her stay at the University, Florence was influenced by two geologists: Charles Van Hise and Roland C. Irving. Despite the gender prejudice she faced for daring to further her education, Florence graduated with two degrees. In 1884, she received her bachelors degree in geology. In 1887 she received her masters in geology.

Bascom spent the next two years teaching natural science at Rockford Seminary in Illinois. About this time, she became interested in a new branch of geology called *petrology*. This new subject was being taught by George Huntington Williams of Johns Hopkins University. In 1890, Bascom went to Johns Hopkins to study under Williams. Unfortunately, Bascom could not officially enroll as a student at Johns Hopkins because of restrictions preventing women from attending the school.

During classes, Bascom was placed behind a screen to remain unnoticed. When she went on field trips, she went alone. Despite these obstacles, Bascom was finally awarded a Ph.D. in geology from Johns Hopkins in 1893. She was the first woman in the United States to receive a Ph.D. in geology and the first woman to receive a Ph.D. in any subject from Johns Hopkins.

Bascom became one of the leading geologists of her time. In 1894, she became the first woman to be elected into the Geological Society of America. At the same time, she taught Geology at Ohio State Universtiy, where

GEOLOGIST

research scientist. Bascom left her teaching post at Ohio State in 1895 and accepted a teaching position at Bryn Mawr College in Pennsylvania. The following year, while still teaching at Bryn Mawr, Bascom became the first woman to serve as assistant geologist on the United States Geological Survey.

In 1906, Bascom became a full professor at Bryn Mawr College. She remained there until her retirement in 1928. During that time

Bascom expanded a single geology course into a full department offering a degree.

The geology program developed by Bascom at Bryn Mawr was extremely successful. In fact, the department became so successful that Bryn Mawr became the major training ground for most women geologists in the 1930s.

In 1945, Florence Bascom died. She is buried beside her family in the Williams College cemetery.

APPLICATION OF SCIENCE

Petrology is the specialized branch of earth science in which the chemical make-up, formation, and weathering of rocks is studied. Petrologists have used their findings to explain how rocks change from one type to another. They have also shown that rocks and the materials that make up rocks continuously cycle through the environment. This process is described in the rock cycle. The diagram shows the changes involved in the rock cycle.

Changes Involved in the Rock Cycle

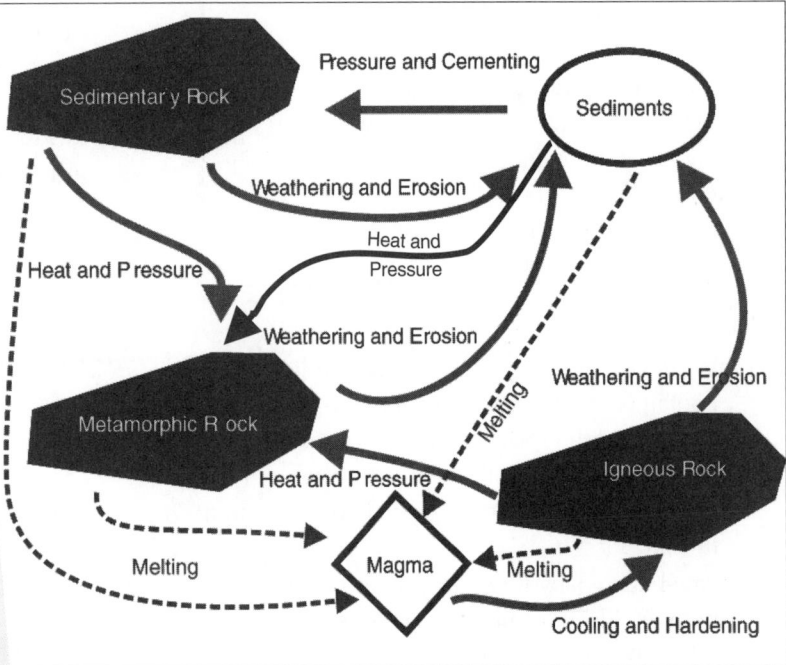

It's YOUR TURN

Hands-On Activity

MECHANICAL AND CHEMICAL WEATHERING

MATERIALS (per group of four)

glass marking pencil, plastic medicine container with lid, water, steel wool, three baby food jars with lids, marble (limestone) chips, vinegar, metric ruler, dropper

SAFETY

Wear safety goggles and a laboratory coat or apron throughout this activity.

BACKGROUND INFORMATION

Weathering is the process by which materials on Earth's surface are broken down and changed in form, either mechanical or chemical. In mechanical weathering, a material, such as a rock, is changed in size and shape. In chemical weathering, the substances making up matter are changed to other substances.

Mechanical weathering is caused by changes in temperature, pressure, and the actions of plant roots or animals. Chemical weathering occurs in the presence of water or moisture in the air.

PROCEDURE

1. Write your group number on the medicine container. Measure the depth of the container and record the measurement in your notes.
2. Fill the container to the top with water and replace the lid. Place the sealed container in a freezer.
3. Label each baby food jar with your group's number. Place a small amount of steel wool into two of the jars. Add just enough water to one jar to cover the bottom of the jar. Place lids on both jars.
4. Observe both jars, noting the color and other traits of the steel wool. Place both jars in an area where they will remain undisturbed.

Observe each jar daily for one week, recording any changes you notice.

5. Place a small amount of marble chips into the third jar. Use the dropper to add four to five drops of vinegar to the chips. Observe what happens and record your observations.
6. After twenty-four hours, remove the container from the freezer. Note and record any changes you observe. Use the ruler to measure the height of the material in the container. Record this measurement.
7. After one week, compare the color and texture of the steel wool samples in each jar you observed. Record your observations in your notes.

ANALYZE AND CONCLUDE

On separate paper, write your answers.

1. Water expands, or increases its volume, by about ten percent when it freezes. How was this illustrated by the water in the medicine bottle?
2. How do you think freezing water may change rocks?
3. After one week, how did the moistened steel wool compare to the dry steel wool?
4. Steel wool contains iron. What can you conclude about the effect of water on the rate at which iron weathers?
5. Vinegar is an acid. What effect does acid have on limestone?

think WORK ACT

CRITICAL THINKING Answer the following questions in complete sentences.

1. What affect do you think acid rain has on materials that contain limestone? Explain.

   _____

   _____

2. Explain how large temperature changes involving freezing and thawing can help break apart rocks.

   _____

   _____

3. Give an example of why it is important for people to know how rocks are changed by mechanical and chemical weathering.

   _____

   _____

GOING FURTHER Complete three of the following.

BUILD YOUR PORTFOLIO

Weathering can corrode metals as well as break down rocks. Survey your community to find examples of weathering. Photograph or list the examples you find.

JOURNAL WRITING

In your journal, describe at least three ways rocks are used by people.

RESEARCH AND REPORT

People have used rocks or stone as a building material for centuries. Research one particular building or monument that is made of stone. Report on when the structure was built, the type of stone used, how long the building process took, and the approximate size of the structure.

COOPERATIVE LEARNING

Rocks are classified as sedimentary, igneous, and metamorphic, depending upon how they formed. Use reference books to find the traits of each type of rock. Make a table of these traits. Then collect at least ten different rocks and classify them using the traits you included in your table.

ALTERNATIVE ASSESSMENT

Metals are often coated with paint or other materials to reduce the effects of weathering. Using an iron nail as your metal, design an experiment to find out how effective different substances are at reducing the effects of weathering on metal. Have your design approved by your teacher. Then carry out your experiment and report on your results.

SYLVIA EARLE MEAD

1935-

Can you remember when you learned to swim? You may recall taking a deep breath, holding your nose, and diving under the water. Your underwater experience lasted only as long as you could hold your breath. Imagine staying under water not for a few seconds, but for two weeks. Such was the experience of Sylvia Earle Mead.

Sylvia Earle was born in New Jersey in 1935. In the 1940s, the Earle family moved to Florida. Marine life in the warm, clear Florida waters sparked Sylvia's interest in the sea. During her senior year in high school, seventeen-year-old Sylvia learned to scuba dive. From 1953 through 1964, Sylvia dived in the Gulf of Mexico and in the waters off North Carolina's coast. When she wasn't diving, Sylvia took classes at St. Petersburg Junior College in Florida. While there, she accepted a job as an assistant in the department of botany at Florida State University. The next year, Sylvia transferred to Florida State as a botany major.

Sylvia Earle received her bachelor's degree in 1955. In 1957, she earned her master's degree in botany from Duke University. She also worked as a fisheries biologist for the United States Fish and Wildlife Service, and taught zoology at St. Petersburg Junior College in 1963 and 1964.

In 1964, Earle joined the International Indian Ocean Expedition. The goal of the expedition was the study of ocean organisms. She was the only woman member of the sixty-member team. During the expedition, Earle collected specimens from the Indian Ocean. The event marked the beginning of Earle's professional and scientific diving career. In future projects, she collected organisms along the islands of the Caribbean Sea, and along the coasts of Panama, Ecuador, Peru, and Chile, and in the waters near the Galápagos Islands.

Earle became research director of the Cape Haze Marine Laboratory in Florida in 1965. The next year, she became director of the facility. She also attended classes at the University of Florida and earned her Ph.D. in 1966.

In 1966, Earle also met Giles W. Mead. He was Curator of Fish at the Harvard University Museum of Comparative Zoology in Boston. The couple later married and Sylvia Earle Mead moved to Massachusetts. She worked as a research scholar at the Radcliffe Institute until 1969.

In 1970, Sylvia Mead was appointed mission leader to the Tektite II project. This project's focus was to learn more about the sea. It also examined the behavior of people living and working in a limited space. During the two-week study, Sylvia Mead and four other women

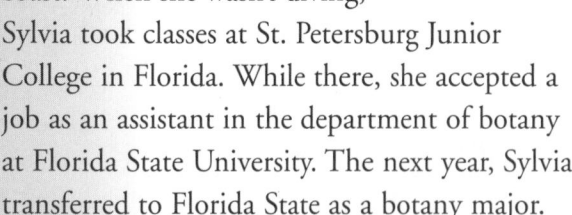

Vocabulary

A **habitat** is the home of an organism.

researchers lived in an underwater habitat fifteen meters (fifty feet) below the surface of the Great Lameshur Bay in the Virgin Islands. The group showed that people could live and work confined in the ocean for extended periods of time. For this work, Mead and her team were invited to the White House. They were also given a Conservation Service Award, the highest award of the United States Department of the Interior.

Following the *Tektite II* project, Mead continued her ocean research. She has served as consultant, chief scientist, and instructor on many underwater projects. Some of these projects were funded by the Smithsonian Institution and the National Geographic Society. Mead has also directed workshops for the National Academy of Sciences and has been a guest speaker at conferences in Sweden and Australia. In 1976, Mead was named a Fellow of the California Academy of Sciences.

BUILDING ON THE PAST

In 1970, Sylvia Earle Mead and four other women scientists began living in an underwater environment as part of the *Tektite II* project. While this was the first such experiment conducted entirely by women, the project built upon knowledge gained in an earlier project called *Tektite I*.

The *Tektite I* project began on February 15, 1969. In this project, a team of four marine scientists lived for two months in a habitat constructed on the ocean floor. Like the *Tektite II* project, the goals of *Tektite I* were to conduct long-term studies of the marine environment and discover how humans reacted to extended stays in an isolated environment. The results showed that people could live and work in such an environment.

Today, similar research into the ways humans react during extended stays in an isolated environment is being conducted on a different frontier—space. The Russian government has successfully placed a space station, called the *Mir*, into orbit around Earth. Thus far, several teams of scientists have journeyed to and worked aboard the space station for periods of time lasting several months. In 1995, history was made when the United States joined in on the research being conducted by sending their astronauts into space to work alongside the Russian scientists.

It's YOUR TURN

Hands-On Activity

THE EFFECT OF DEPTH ON PRESSURE

MATERIALS (per group of two)

empty cardboard quart-size milk container, safety pin, large bin, water, metric ruler, cellophane tape

SAFETY

Clean up spills immediately. Work carefully with the safety pin. Wear a laboratory coat or apron and safety goggles throughout the activity.

BACKGROUND INFORMATION

Water pressure is a serious consideration in diving or other prolonged stays underwater. Pressure is a measure of mass per unit area. As one goes deeper in water, pressure increases because the amount of water above you increases. As the amount of water increases, the weight of the water pressing down on you also increases. Knowledge of how pressure changes with depth is important in the design and construction of submarines and other underwater exploration vehicles or habitats.

PROCEDURE

1. Using the ruler as a guide, make a mark down the center of one side of the milk container from top to bottom.
2. Place the ruler along the line you drew. Starting at the bottom of the container, use your safety pin to make a hole in the container every three centimeters from the bottom to the top. Make sure each hole goes completely through the container.
3. Cover the holes using a strip of cellophane tape.
4. Fill the container to the very top with water.
5. Place the container into a collecting bin. Quickly remove the strip of tape from the container. Observe the flow of water from each hole. Make a diagram of your observations.

ANALYZE AND CONCLUDE

On separate paper, write your answers.

1. How many holes did you make in the container?
2. Describe the flow of water coming out of the top hole.
3. How does the flow of water coming from the center hole compare with that coming from the top hole?
4. How does the flow of water coming from the bottom hole in the container compare to that coming from the center hole?
5. Based upon your observations, where do you think the greatest pressure existed in the container?
6. Why is the pressure greatest at this part of the container?
7. What is the effect of water depth on pressure?

think WORK ACT

CRITICAL THINKING Answer the following questions in complete sentences.

1. How might Sylvia Earle Mead's training as a botanist have helped her in her career as a marine biologist?

   _____

   _____

2. What might be some difficulties and dangers faced by researchers living in a sealed environment in the ocean?

   _____

   _____

3. Name two other environments that are confined and under pressure.

   _____

GOING FURTHER Complete three of the following.

BUILD YOUR PORTFOLIO

Do library research on either the *Tektite I* or *Tektite II* habitat. Make a drawing that shows what the habitat looks like and where people live and work in the habitat. Label your drawing.

JOURNAL WRITING

If you had an opportunity to live in a sealed environment beneath the ocean's surface for one year, would you go? Why or why not?

RESEARCH AND REPORT

People have begun living in space aboard space stations. Find out how a space station meets the needs of the people living there. Explain how this technology might be applied to people living in the ocean.

COOPERATIVE LEARNING

In your group, develop a model for an underwater ocean environment in which people could live for long periods of time. Show where people will work and where they will live. Explain how food and oxygen will be supplied and how people can safely enter and exit the environment.

ALTERNATIVE ASSESSMENT

Find out how the needs of the fishes raised in a saltwater aquarium differs from that of a freshwater aquarium. Set up both types of aquariums and keep track of all the variables you need to maintain in each for a period of three months. Write a report of your results explaining how you maintained each environment.

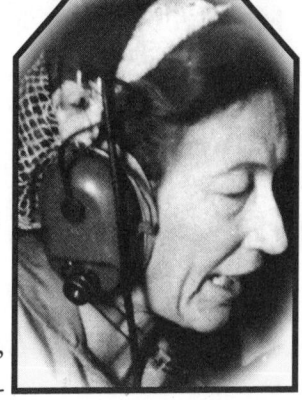

Meteorologists study weather and conditions of the atmosphere. Weather is determined by air pressure, temperature, wind speed and direction, humidity, and other factors. Many meteorologists study and predict the weather. Others actually try to control or change the weather. One scientist who has experimented with changing the weather is Joanne Malkus Simpson.

Joanne Malkus was born in Boston on March 23, 1923. Her father was a newspaper editor; her mother was a newspaper reporter. Throughout her youth, Malkus went to private schools. She did well in school, especially in science and mathematics.

After graduating high school in 1940, Malkus went to the University of Chicago. During her first year of college, the United States became involved in World War II. Malkus volunteered to train military personnel about weather patterns and conditions, especially those that affect pilots. As a child, Malkus had made many airplane flights with her father. Her personal experiences as well as courses she had taken contributed to her expertise in meteorology. While teaching the military personnel, Malkus contin-

ued working toward her bachelor's degree in meterology, which she received in 1943. She was the first woman in the United States to receive a degree in meteorology.

Malkus continued her education. She received her master's degree in 1945 and her Ph.D. in 1949. At the same time, she taught meteorology at the college level. As a woman working in meteorology, Malkus was faced with constant prejudice and discouragement.

Malkus wanted both a career and a family, but she was not always successful in achieving these goals. Malkus had twice married co-workers. In each case, she was fired because the universities did not allow employment of both husbands and wives. As was the custom, the wife was released from her position. In addition to losing her jobs, another personal setback was that both marriages ended in divorce. Her third marriage to Robert Simpson has endured.

Joanne Malkus Simpson has held many positions. From 1946 to 1949, she was an instructor and later professor of physics and meteorology at the Illinois Institute of Technology. From 1951 to 1961, she served as a meteorologist at the Woods Hole Oceanographic Institute, a marine research facility in Massachusetts. She later accepted a position as a professor of meteorology at the University of California at Los Angeles. While at the University of California, Simpson also

> ## Vocabulary
> Seeding clouds means injecting them with chemicals to induce rain

METEOROLOGIST

served as a consultant for the U.S. Weather Bureau and conducted original research. One such research project involved changing the weather. Using crystals of silver iodide, Simpson was successful in seeding clouds to make rain. The seeding was done to lessen the strength of storms and change their behavior.

After leaving the University of California, Simpson became director of the U.S. Weather Bureau's Experimental Meteorological Laboratory of the National Oceanic and Atmospheric Administration (NOAA) in Coral Gables, Florida. She held this position from 1965 to 1973. She was also an adjunct professor of Atmospheric Sciences at the University of Miami. Later, she became professor of environmental sciences at the University of Virginia.

Joanne Malkus Simpson is a member of many national committees as well as England's Royal Meteorological Society. In 1963, the *Los Angeles Times* named her Woman of the Year, and she has been nominated three times by the *Ladies Home Journal* for the honor of Woman of the Year in Science. She continues to encourage women to enter the field of meteorology.

BUILDING ON THE PAST

During its early stages, weather forecasting was based upon data collected from farmers, local weather stations, hot-air balloons, and reports from pilots. Over time, the system of weather analysis and forecasting has improved. Although many former methods and techniques of weather study and forecasting are still used, technology has made the study of weather more accurate.

Today, high-altitude planes, Doppler radar systems, weather satellites, and computers are used in the study of weather. Often, these technologies are used in combination with one another to track weather systems and forecast weather conditions. For example, as weather systems move across the country, picking up or releasing moisture, meteorologists can more accurately predict and prepare communities for changing weather conditions. Often, as with the case of severe storms and hurricanes, these predictions save both lives and property.

It's **YOUR TURN**

Joanne Malkus
SIMPSON

Hands-On Activity

CONDENSATION OF WATER VAPOR

MATERIALS

(per group of two)

hot tap water, ice cubes, baby food jar,
4″ x 4″ glass plate

SAFETY

Wear goggles and a laboratory coat or apron throughout this activity. Clean up any spills that occur immediately. Be careful when working with hot tap water and glass.

BACKGROUND MATERIAL

When a substance changes from a solid to a liquid, a liquid to a gas, or from a solid to a gas, it cools its surroundings. Air acts like a sponge to absorb water in the form of a gas. This gas is called *water vapor*. This property of air is affected by temperature. Warm air can hold more water than cool air. If air is cooled enough, it releases water vapor as liquid water. This principle is used to make rain artificially.

PROCEDURE

1. Fill a baby food jar one-quarter full of hot tap water.
2. Place the glass plate over the mouth of the baby food jar. Place about two ice cubes on top of the glass plate.
3. After several minutes, observe the inside of the jar and the bottom of the glass plate. Record your observations.

ANALYZE AND CONCLUDE

On separate paper, write your answers.
1. What formed on the bottom of the glass plate after the ice was added?
2. What do you think caused this to occur?
3. What effect did the ice have on the water vapor in the jar?
4. What is this process called?

think WORK ACT

CRITICAL THINKING Answer the following questions in complete sentences.

1. Why would pilots need to be knowledgeable about weather conditions?

    _____

    _____

2. Why might scientists be interested in seeding clouds to produce rain?

    _____

    _____

3. Joanne Malkus Simpson was forced to leave several jobs because her husbands worked for the same employer. Why do you think an employer might not want husbands and wives to work together? Do you agree or disagree with this thinking? Why?

    _____

    _____

GOING FURTHER Complete three of the following.

BUILD YOUR PORTFOLIO

Do library research to identify the characteristics of different cloud types. Make a drawing or a model that compares the appearance and heights of the different types of clouds.

CONCEPT MAPPING

Find out how precipitation, condensation, and evaporation are involved in the water cycle. Make a concept map of your findings.

RESEARCH AND REPORT

Do library research to find out what absolute humidity and relative humidity are and how they are measured. Present an oral report of your findings to the class.

COOPERATIVE LEARNING

With your group, set up a weather station at your school. Your station should measure temperature, rainfall, humidity, air pressure, wind speed, and wind direction. Obtain or design the devices needed to make these measurements. Record these data twice each day for one month. Organize your findings in a table or a graph.

JOURNAL WRITING

In your journal, describe three ways that weather is important to your everyday life.

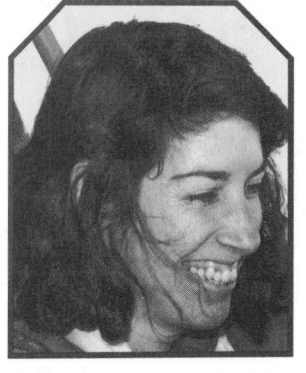

The conquest of space began long before people went to the moon. For some, it began with the telescopic observations of Galileo, the 17th century Italian physicist and astronomer. For others, it began with the fictional writings of French novelist, Jules Verne. In his book, *From the Earth to the Moon*, Verne told an imaginative adventure story about a spaceship and space travel. Ellen Ochoa's conquest of space began in 1987 when she learned she had been selected by NASA to become an astronaut.

Ellen Ochoa was born in Los Angeles, California, in 1958. She was one of five children. While in junior high school, Ellen's parents divorced. Ellen moved with her mother, Rosanne, to La Mesa, a suburb of San Diego. There, Ellen attended school and prepared for her future.

Ochoa's mother delayed her own dream of going to college to take care of her family. Her mother was an inspiration and a role model for Ochoa. In 1959, one year after Ellen's birth, Rosanne Ochoa enrolled at San Diego State University. She spent the next twenty-three years pursuing a bachelor's degree. Her determination paid off in 1982 when she was awarded

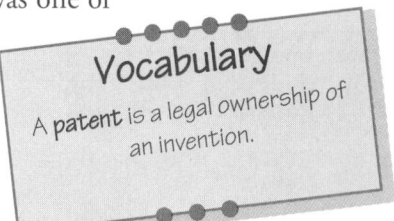

Vocabulary

A **patent** is a legal ownership of an invention.

a bachelor's degree with a triple major in business, biology, and journalism.

Like her mother, Ellen believed in the value of a good education. While in junior high school, she excelled in math and science. Later, Ellen graduated at the top of her class from Grossmont High School in the 1970s. She repeated this achievement when she graduated first in her class from San Diego State University with a degree in physics in 1980.

After earning her bachelor's degree, Ellen went to Stanford University. She earned her master's degree in electrical engineering in 1981. In 1985, she earned a doctorate in electrical engineering. That same year, she applied for admission into the NASA space program.

From 1985 to 1988, Ochoa worked as a research engineer at Sandia National Laboratories in California. She also worked in the Intelligent Systems Technology Branch at NASA/Ames Research Division. Additionally, Ochoa holds several patents in optical processing.

While working as a research engineer, Ochoa was chosen by NASA to train as an astronaut. Her training began at the Johnson Space Center in 1990. Upon completing her training, Ochoa became the first Latin American female astronaut. Three years later, Ochoa became the first Latin American woman in space during a nine-day mission aboard the shuttle *Discovery*.

As a result of her work as both an engineer and an astronaut, Ochoa has received many

ASTRONAUT ENGINEER

awards. Among them is the Hispanic Engineer National Achievement Award for Most Promising Engineer in Government. She was presented with this award in 1989. The following year, she received the Pride Award from the National Hispanic Quincentennial Commission.

Ochoa is also noted for her public speaking appearances. She has stated that being a role model is one of her top priorities. She believes strongly in education and takes great pride in her accomplishments. Ochoa is hopeful that young Latin American children will see a bit of themselves in her.

BUILDING ON THE PAST

NASA selected its first group of astronauts in 1959. The team, known as the Mercury astronauts, was made up of seven men: M. Scott Carpenter, L. Gordon Cooper, John Glenn, Virgil Grissom, Walter Schirra, Alan Shepard, and Donald Slayton. By 1965, each of these men, except Slayton, had participated in at least one space mission.

Between 1962 and 1969, NASA assembled six more astronaut groups. The six groups included a total of sixty-five men. It was not until the selection of astronaut Group 8 in 1978 that women were chosen for training in the astronaut program. Group 8 was made up of a total of thiry-five astronauts; five of them were women. Since Group 8, all of the astronaut groups assembled by NASA have included women, as shown in the timeline.

1978 Group 8: Of the 35 astronauts selected, 5 were women. The women were: Anna Fisher, Shannon Lucid, Judith Resnick, Sally Ride, Margaret Seddon, and Kathryn Sullivan.

1980 Group 9: Of the 19 astronauts, 2 were women. The women were Mary L. Cleave and Bonnie J. Dunbar.

1981: On April 12, the space shuttle program begins with the launch of *Columbia*.

1984 Group 10: Of the 17 astronauts, 3 were women. They included Ellen Baker, Marsha Ivins, and Kathryn Thornton.

1985 Group 11: Of the 13 astronauts, 2 were women: Linda Godwin and Tamara Jernigan.

1987 Group 12: Of the 15 astronauts, 2—N. Jan Davis and Mae C. Jemison—were women.

1990 Group 13: Of the 23 astronauts, 4 were women. The group included: Eileen Collins, Susan Helms, Janice Voss, and Ellen Ochoa.

It's YOUR TURN

Ellen OCHOA

Hands-On Activity

COMPARING IMAGES FORMED BY LENSES

MATERIALS

(per group of two)
convex lens, concave lens

BACKGROUND INFORMATION

A lens is a curved, transparent material that allows light to pass freely through it to produce an image. There are two main types of lenses: convex lenses and concave lenses. A convex lens is one that is thicker at its center than at its edges. A concave lens is thinner at the center than at its edges. Each type of lens forms a different type of image.

PROCEDURE

1. Using your convex lens, view the Background Information text on this page. Move the lens toward and away from the page until a clear image forms. Note if there are any changes in the image as you move the lens.
2. Repeat step 1 using the concave lens.

ANALYZE AND CONCLUDE

On separate paper, write your answers.

1. Describe the shape of a convex lens.
2. Describe the shape of a concave lens.
3. Describe the image formed by the convex lens.
4. Describe the image formed by the concave lens.
5. Based upon your observations, what kinds of instruments might use convex lenses?
6. What kinds of instruments might use concave lenses?
7. Which of these two lenses is used in a hand lens?

think WORK ACT

CRITICAL THINKING Answer the following questions in complete sentences.

1. How might a degree in engineering help with a career as an astronaut?

   _____

   _____

2. Ellen Ochoa has stated that being a role model is one of her top priorities. What traits of Ochoa help her to serve as a role model?

   _____

   _____

3. Ellen Ochoa s mother began college the year after Ellen was born. She received her bachelor s degree twenty-three years later. What impact do you think this had on Ellen as a student?

   _____

   _____

GOING FURTHER Complete three of the following.

BUILD YOUR PORTFOLIO

Write to or E-mail NASA to get information about women in the space program. Use the information you obtain to create a scrapbook. Place a copy of the letter you sent to NASA and your scrapbook in your portfolio.

PERFECT YOUR SKILL

Use the data from the timeline on p. 85 to make a graph showing the percentages of men and women making up each astronaut group.

RESEARCH AND REPORT

Read Jules Verne's *From the Earth to the Moon.* Write a brief summary of the novel. Describe one way in which the novel accurately depicts space travel and one way in which the description of space travel is completely fictional.

COOPERATIVE LEARNING

Select one or two astronaut groups from the time-line on p. 85. Divide the names so each person researches the work of one astronaut. Combine the information gathered in a chart or some other visual device.

JOURNAL WRITING

In your journal, describe ways in which becoming an astronaut interests you. What skills do you think you would need to improve to be an astronaut.

HENRIETTA SWAN LEAVITT

1868-1921

Stars are masses of very hot gases. Individual stars differ in size, distance from Earth, and temperature. Star color also varies with temperature. Red stars have the lowest temperatures; blue-white stars are the hottest. All these traits determine how bright a star appears to people on Earth. The term *magnitude* is used to describe star brightness. Henrietta Swan Leavitt spent much of her life's work measuring star magnitudes.

Henrietta Swan Leavitt was born in Cambridge, Massachusetts, in 1868. She was one of seven children born to George Leavitt, a minister, and his wife, Henrietta. Young Henrietta was seriously hearing impaired; however, her hearing problems did not prevent her from doing well in school.

The family moved to Cleveland, Ohio, where Henrietta attended Oberlin College. After three years at Oberlin, Henrietta returned to Cambridge to finish her education at Radcliffe College. At the time, Radcliffe was known as the Society for the Collegiate Instruction of Women. During her last year at Radcliffe, Leavitt took an astronomy course that sparked

her interest in stars. After graduating in 1892, she stayed at Radcliffe another year to take more astronomy courses.

From 1893 to 1895, Leavitt took time off to travel. She then returned to Cambridge, where she did volunteer work at the Harvard Observatory. Other noted women astronomers, such as Annie Jump Cannon and Williamina Fleming, also worked at the Harvard Observatory.

Edward Pickering was director of the Harvard Observatory. Pickering was unusual in his open-mindedness and fairness toward women astronomers. Unlike many male scientists at the time, Pickering encouraged the work of women scientists and treated them as equals.

Leavitt was a meticulous worker. In 1902, she became a permanent member of the observatory staff. Her job was to determine the magnitudes of stars from their photographs. Photographic plates are more sensitive and show many more wavelengths of light than humans can see; but identifying the different wavelengths on them is a detailed, tedious task.

Leavitt's most important invention involved variable stars. These are stars that brighten and dim at regular intervals. In 1908, Leavitt reported her belief that there was a relationship between the length of a variable star's period of brightness and its magnitude. In 1912, she made a table which showed that the longer the interval of variation, the brighter the star. This

> ## Vocabulary
>
> Magnitude is a measure of a star's brightness.

table is known as the *Period Luminosity* Relationship for *Variable Stars* and is still used. Over time, the table became a basic tool of astronomers who used it to find the distances of stars in the Milky Way Galaxy and other galaxies.

In 1913, Leavitt's system for finding the colors of north polar stars was adopted by the International Committee on Photographic Magnitudes. Now Leavitt's method is used to find the colors of stars in other parts of the sky.

Leavitt was a member of the Astronomical and Astrophysical Society of America. She worked at the Harvard Observatory her entire career. In total, she discovered 2400 variable stars, about half the known variable stars at the time. She also discovered four novas. A nova is a star that glows brightly, then dimly. She died of cancer at age fifty-two.

APPLICATION OF THE SCIENCE

The actual or absolute magnitude of a star is determined by its size and temperature. Once the absolute magnitude of a star is determined, the distance of the star from Earth can also be determined. This distance is measured in units called light-years. A light-year is the distance that light travels in one year. This distance is equal to 9.5 trillion kilometers.

In addition to light years, scientists have developed other units for measuring distances in space. For example, the astronomical unit, (AU) is used to measure the distances within our solar system. One AU is equal to 93 million miles (1.5 million kilometers). This is the average distance from Earth to the sun.

A parsec is used to measure even larger distances, such as those between distant stars and galaxies. One parsec is equal to 3.2 light years.

© 1996 The Peoples Publishing Group, Inc.   **Multicultural Women of Science ◆ Henrietta Swan Leavitt**

It's YOUR TURN

Hands-On Activity

MAKING A SIMPLE LIGHT METER

MATERIALS (per individual)

Six pre-marked paper disks provided by your teacher, safety scissors, stapler, three lamps of different intensities

SAFETY

Be careful when working with the scissors and light sources. Do not look directly into the light sources.

BACKGROUND INFORMATION

The lower the magnitude of a star, the brighter the star. Astronomers use light meters and other tools to determine the magnitude (brightness) of a star. The ability of light to pass through materials can also be used to measure brightness.

PROCEDURE

1. Cut out section 1 from the first disk. Cut out sections 1 and 2 from the second disk. Cut out sections 1 to 3 from the third disk. Repeat this sequence for each of the remaining disks, except the last. Leave this disk intact.
2. Place the intact disk on your desktop. Place the disk with only space 1 removed on top of the intact disk so that space 1 of the intact disk remains uncovered.
3. Place the disk with two spaces removed on top of the disks from step 2 so that spaces 1 and 2 remain uncovered. Repeat this sequence for each of the remaining disks. Once all the disks are piled together, staple them through the center.
4. Hold your light meter up to a light source. Note the various brightnesses of light that come through the varying thicknesses of paper.
5. Use your meter to determine the intensity of the three light sources identified by your teacher. For each light source, record the highest section number through which light can still be seen.

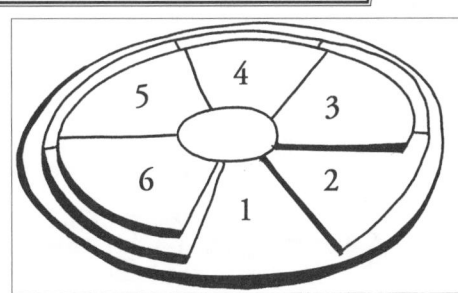

Paper Disk Light Meter

ANALYZE AND CONCLUDE

On separate paper, write your answers.

1. Which number on your meter determines the brightest light? How many thicknesses of paper does this number represent?
2. Which number on your meter determines the dimmest light? How many thicknesses of paper does this number represent?
3. Of the three lights observed, which was brightest? Next brightest? Dimmest?
4. How did the scale you developed to measure light magnitude with your meter differ from that used by astronomers to measure star magnitudes?

think WORK ACT

Henrietta
SWAN LEAVITT

CRITICAL THINKING Answer the following questions in complete sentences.

1. How does the magnitude of variable stars differ from those of other stars?

   _____

   _____

2. Which variable star would be brighter, one that takes a long period of time to change its brightness or one that changes its brightness over a shorter period of time?

   _____

   _____

3. Betelgeuse is a large red star. Rigel is a large blue-white star. Which of the two stars has a higher temperature? Explain.

   _____

GOING FURTHER Complete three of the following.

BUILD YOUR PORTFOLIO

Observe and sketch the night sky. Use star charts and library sources to identify any constellations you have sketched and the names of the major stars making up each constellation.

PERFECT YOUR SKILL

Use library resources to find the magnitudes of the following stars: Sun, Algol, Proxima centauri, Vega, Deneb, Polaris, Mizar, Betelgeuse, Rigel, and Aldebaran. Place the stars in order from brightest to dimmest.

JOURNAL WRITING

Select and draw a star group in the night sky. Name your star group and write a short story about it explaining the significance of its name.

CONCEPT MAPPING

Use a dictionary to find the difference in meaning between the terms *luminous* and *illuminated*. Develop a concept map that explains the differences in these terms and place the following astronomical bodies beneath the correct heading: Venus, sun, moon, planetoids, meteors, and Earth.

RESEARCH AND REPORT

Choose a constellation from a star chart. Do library research to find out how the constellation was named and what stories or superstitions are connected with the constellation. Present an oral report of your findings.

ANNIE JUMP CANNON
1863-1941

Stars have been a source of wonder and curiosity for thousands of years. Ancient mariners found their way using the fixed position of Polaris—the North Star. Early peoples imagined patterns formed by groups of stars, and named the constellations. More recently, people have had stars named after them. As a young girl, Annie Jump Cannon was also fascinated with stars. She made the study of stars her life's work.

Annie Jump Cannon was born in Dover, Delaware. Her father, Wilson Lee Cannon, was a shipbuilder and a member of the Delaware senate. Her mother, Mary Elizabeth Cannon, had a strong interest in stars. As a young girl, Mary had taken a course in astronomy. Later, Mary and her daughter Annie studied the stars and constellations from the attic of their home.

Annie Jump Cannon attended public school. She then entered the Wilmington Conference Academy in Dover. She graduated at age sixteen and entered Wellesley, a new women's college in Massachusetts. While at Wellesley, Cannon developed an interest in spectroscopy.

Cannon graduated from Wellesley in 1884.

Vocabulary

Spectroscopy is the study of how light breaks apart after it passes through a prism.

For the next ten years, she traveled, studied music, and improved her photography skills. Following the death of her mother in 1893, Cannon's childhood interest in stars was renewed. She returned to Wellesley in 1894 to take graduate classes. The next year, she moved to Cambridge to study astronomy, physics, and mathematics at Radcliffe College.

In 1896, Cannon joined the staff of the Harvard University Observatory in Boston. The observatory was known for its work on the spectral classification of stars. Her job was to analyze photographic plates and interpret the spectra given off by stars. Cannon enjoyed her work at the observatory, but during the cold winters, she often suffered from severe colds and sinus infections. As a result of these infections, Cannon eventually lost her hearing.

For the quality of her work, Annie Jump Cannon is referred to as the most famous woman astronomer in the world. For the amount of work completed, she is referred to as the "census taker of the sky." Between 1918 and 1924, Cannon published nine volumes of her work. They included the classification of more than a quarter million stars. In total, Cannon classified nearly one-half million stars. She also discovered five novas—stars that become extremely bright then fade—and 300 variable stars—stars whose brightness appears to change. In 1922, her system of classification was adopted by the by the International Astronomical Union. Today, her classification scheme is the official system for the spectral classification of stars.

For her accomplishments, Cannon received many honors and awards. She was the first woman to receive an honorary Doctor of Astronomy degree from the University of Gröningen, in the Netherlands, and an honorary Doctor of Science degree from Oxford University in England. She was also made an honorary member of the Royal Astronomical Society. Annie Jump Cannon retired in 1940. She died the following year at age seventy-eight.

APPLICATION OF THE SCIENCE

Stars are great distances from Earth, yet scientists have been able to learn much about them. How is this accomplished? Much of what is known about individual stars has been determined from the light given off by the stars. A star is a large mass of gases that gives off extreme amounts of heat and light. In a technique known as spectroscopy, the light given off by a star is broken down into a spectrum using a prism. Elements that are heated to high temperatures give off different colors of light. The elements that make up a star can be identified by the spectra, or colors of light, they produce.

The lines making up the spectrum of a star may shift toward or away from a particular color of the spectrum. This indicates in which direction (toward or away) a star is moving in relation to the Earth. It also provides information about the speed at which a star is moving. The colors given off by stars also provide information about the temperature of the star, its distance from Earth, and its size.

It's YOUR TURN

Hands-On Activity

ANALYZING SPECTRA

MATERIALS (per student)

white unruled paper, diffraction grating spectroscope, 1 piece of red cellophane, 1 piece of blue cellophane, colored pencils or crayons

MATERIALS (per class)

incandescent light source, fluorescent light source, neon light source, sodium light source, mercury light source (if available)

SAFETY

Never look at the sun or any high-intensity light source.

BACKGROUND INFORMATION

Light passing through a prism or diffraction grating is broken up into colors of the spectrum. Any body that gives off its own light can be studied using a diffraction grating or prism. An instrument that uses a prism to study the makeup of light is called a spectroscope. The spectra of stars and other luminous bodies can be used to identify the chemical makeup of the body.

PROCEDURE

1. Observe a section of bright sky through the slit in your spectroscope. Do not look directly at the sun. Rotate the tube until the light entering the tube is broken apart into a series of colors. Using the colored pencils, sketch what you see on the sheet of white paper.
2. Hold a piece of red cellophane over the slit. Sketch what you see.
3. Repeat step 2 using the blue cellophane.
4. Hold both the red and blue cellophane over the slit. Sketch what you see.
5. Observe the spectrum of incandescent light. Sketch what you see. Repeat this procedure for the fluorescent light.
6. Observe the spectra of each of the other light sources provided by your teacher. Sketch the spectrum of each light source. Label each sketch with the type of light source.

ANALYZE AND CONCLUDE

On separate paper, write your answers.

1. When you looked at the bright sky, you were observing the solar (sun s) spectrum. How many color zones did you observe? How did the zones compare in size?
2. What effect did the blue filter have on the solar spectrum you observed? The red filter?
3. What effect in general does the use of a color filter have upon its corresponding color in the produced spectrum?
4. What similarities and differences did you see in the other spectra you observed? What do you think causes these differences?

think WORK ACT

CRITICAL THINKING Answer the following questions in complete sentences.

1. Why do you think the spectrum of a light source can be compared to a fingerprint?

   _____

   _____

2. Why is Annie Jump Cannon sometimes referred to as the "census taker of the sky"?

   _____

   _____

3. How were Annie Jump Cannon's interests as a child and young girl related to her career choice?

   _____

   _____

GOING FURTHER Complete three of the following.

BUILD YOUR PORTFOLIO

Observe the sky on a clear night. Try to make your observations from a location where there are not a lot of artificial light sources, such as street lights. Make a drawing that shows the approximate positions of several stars. Indicate directions, such as north and south, on your drawing for reference. Use library sources to identify any constellations or star groups that appear in your drawing.

PERFECT YOUR SKILL

Use the spectroscope from the activity to determine what effect brightness or magnitude of light has on the spectrum produced by a light source. Make drawings of the spectra produced by incandescent and fluorescent bulbs of different wattages. Write a report of your findings.

COOPERATIVE LEARNING

Find at least five examples of paintings, songs, or poems that have stars as their themes. Explain how stars are important to each work. Decide if the stars are portrayed in an accurate manner or if they are used only for setting. Explain.

RESEARCH AND REPORT

Use library resources to find out how star spectra are used to identify the elements present in stars. Write a report of your findings that includes drawings of the spectra produced by several elements.

JOURNAL WRITING

In your journal, write a poem, short story, or song about stars.

M^{AE} JEMISON
1956 -

On September 12, 1992, a fiery blast, a huge cloud of steam, and a thunderous roar launched the space shuttle *Endeavor* into an eight-day mission around Earth. Seven astronauts were aboard, including Dr. Mae Carol Jemison, a science mission-specialist. That day, Jemison became the first African American woman astronaut to go into space.

Mae Jemison was born in Decatur, Alabama, in 1956. She was the youngest of three children. Her father, Charlie Jemison, was a carpenter, and a maintenance supervisor. Her mother was an elementary school teacher.

Although she was born in the South, Jemison was raised in Chicago. As a child, young Mae looked at the stars, believing she would someday travel in space. At the time, she was not aware that space travel was reserved only for male jet pilots. In 1969, the successful *Apollo 11* moon mission strengthened Jemison's desire to fulfill her dream.

While in high school, Jemison's interest in space heightened. She developed a sound scientific base by reading books on astronomy and space travel. Her interest spurred her to make frequent visits to the Museum of Science and Industry, learning more with each visit.

> **Vocabulary**
> **Astronomy** is the science of space and celestial objects.

After high school, Jemison entered Stanford University as a National Achievement Scholar. She started with a double major—chemical engineering and African and Afro-American studies. In 1977, she received her bachelor's degree in chemical engineering. She then entered Cornell University Medical College in Ithaca, New York.

While at Cornell, Jemison was president of the Cornell Medical Student Executive Council and the Cornell Chapter of the National Student Medical Association. Her membership in this second group gave her an opportunity to travel and study in Thailand, Cuba, and Kenya. Jemison received her medical degree in 1981. The following year, she became a general practitioner in Los Angeles.

In 1983, Jemison left her medical pratice and joined the Peace Corps as a medical officer. For the next two years, she worked as a physician in West Africa. When she returned to the United States in 1985, she took action to realize her dream of space travel. She applied for admission into the NASA space program. She also took night school classes in engineering at the University of California in Los Angeles.

In January 1986, the tragic explosion of the *Challenger* stalled America's space program temporarily; however, Jemison remained determined to become an astronaut. In October 1986, she applied once again for admission to

ASTRONAUT
PHYSICIAN—
CHEMICAL ENGINEER

NASA. Less than a year later, she became the first African American woman to be admitted into the NASA space program. She completed her training and became an astronaut in the summer of 1988. She was named science mission-specialist. Her job was to perform scientific experiments on conditions in space during a shuttle mission.

On September 12, 1992, the space shuttle *Endeavor* roared off its pad at the Kennedy Space Center in Florida. Dr. Jemison's job on the mission was to perform scientific experiments on motion sickness, calcium loss in bones, and weightlessness. Her experiments generated much scientific data.

Jemison resigned from NASA in 1993 to start the Jemison Group, a technology firm in Houston, Texas. The company researches and produces advanced technology, like that used in space travel, for use in developing nations. Specifically, the Jemison Group focuses on improving communications and health care in West Africa. Mae Jemison is "giving back" her talents to the world.

IMPLICATION OF THE SCIENCE

The products and processes that result from aerospace research are called *spinoffs*. Examples of spinoffs include the heart pacemaker, the material known as mylar, and the development of laser surgery.

Laser angioplasty is used by doctors to clear blocked coronary arteries without surgery. This procedure makes use of a device called a "cool" laser. The cool laser process was designed by NASA to measure gases in Earth's atmosphere.

Heart imaging systems are another example of a spinoff. NASA developed this technology to survey and study Earth resources. The heart imaging system uses television like monitors to show procedures being performed on the heart. This technology allows cardiac procedures to be performed more accurately and with less risk than surgery.

One of the most exciting developments of space technology is the concept of virtual reality. Virtual reality is an imaginary environment that can be entered or interacted with by a human. NASA's interest in virtual reality stems from the need for telesurgery—surgery performed by a robot, and for the large-scale control of robotic space systems. On Earth, virtual reality systems are being explored largely as a form of entertainment.

It's YOUR TURN

Mae JEMISON

Hands-On Activity

THE EFFECT OF CALCIUM IN BONE

MATERIALS

(per group of two)

large, mayonnaise-type jar with lid, glass marking pencil, vinegar, clean chicken bone, tongs

SAFETY

Wear goggles and a laboratory coat or apron throughout this activity. Clean up any spills that occur immediately.

BACKGROUND INFORMATION

Calcium is an important component of bones and teeth. Usually, this calcium is in the form of calcium carbonate or calcium phosphate. The calcium in bone is mostly calcium phosphate.

PROCEDURE

1. Use the glass marking pencil to label the jar with your group name and number.
2. Fill the jar two-thirds full with vinegar.
3. Note the hardness of the bone. Using the tongs, place the bone into the vinegar in the jar. Place the lid on the jar.
4. Store the jar as directed by your teacher.
5. After three days, use the tongs to remove the bone from the vinegar. Rinse the bone thoroughly with tap water. Again note the hardness of the bone. Use the tongs to place the bone back in the jar.
6. In three to five days, repeat step 5. After you have observed the bone, discard it as directed by your teacher.

ANALYZE AND CONCLUDE

On separate paper, write your answer.

1. How did the hardness of the bone in step 5 compare with the hardness of the bone in step 3?
2. How did the hardness of the bone in step 6 compare with the hardness of the bone in step 5?
3. What do you think caused the change in the bone?
4. What substance in the bone did the vinegar dissolve or break down?

Name _____   Date _____

think WORK ACT

CRITICAL THINKING Answer the following questions in complete sentences.

1. On her space mission, Jemison studied calcium loss in bones. How might a loss of calcium affect you?

   _____

   _____

2. Why would it be important to have a science mission specialist aboard a space flight?

   _____

   _____

3. The United States space program costs millions of dollars each year. Many people have suggested that space exploration by the government be stopped to save tax dollars. Explain your views on this topic.

   _____

GOING FURTHER Complete three of the following.

BUILD YOUR PORTFOLIO

Write to NASA requesting information about a specific astronaut. Include a copy of your letter and any information you receive in your portfolio.

PERFECT YOUR SKILL

Devise an experiment to determine how much food and oxygen you would need to complete a four-day space mission.

RESEARCH AND REPORT

Obtain a copy of the 1990 government publication, *Spinofff.* Choose one material in the publication to learn more about. Present an oral report of the importance of the material you researched to your class.

COOPERATIVE LEARNING

Think about the condition of weightlessness in space. Taking this condition into account, prepare and design packaging for three foods you would take into space.

JOURNAL WRITING

Assume you are going to interview Mae Jemison. Write five to seven questions in your journal that you would most like to ask her.

WOMEN OF

CHEMISTRY

MARIE CURIE
1867-1934

DOROTHY HODGKIN
1910-1994

IRÉNE JOLIOT-CURIE
1897-1956

MARIA GOEPPERT-MAYER
1906-1972

AND

PHYSICS

LISE MEITNER
1878-1968

ROSALYN YALOW
1921-

MARIE CURIE
1867-1934

If you were asked to name a famous woman scientist, who would you choose? Most people would probably name Marie Curie. Marie Curie was the first and the second woman to win a Nobel Prize in science. She also inspired many others to work in science, including several Nobel Prize winners.

Marie Curie was born Marya Skodowska in Warsaw, Poland, in 1867. In her youth, Marya did well in school. She graduated from high school at age fifteen. At that time, women in Poland were not permitted to attend college. Marya moved to Paris. In 1891, she enrolled at the Sorbonne, a part of the University of Paris. It was at this time that Marya changed her name to Marie, the French form of Marya.

In only two years, Marie earned the equivalent of a master's degree in physics from the Sorbonne. The next year, she earned a similar degree in mathematics. At the Sorbonne, Marie also met Pierre Curie, a physicist who was studying crystals and magnetic materials. In 1895, Marie planned to return to Poland. Her plans changed when Pierre proposed to her. The couple married and settled in Paris.

> ### Vocabulary
> An **element** is a substance that cannot be broken down chemically. There are more than 100 known elements.

In 1897, Marie Curie began working on her doctorate. She decided to study the energy given off by the element uranium. Pierre soon abandoned his work with crystals and magnetic materials to join his wife in her studies. Together, the Curies discovered that the energy given off by uranium came from within its atoms. Marie coined the term *radioactivity* to describe this phenomenon. For their work on radioactivity, Marie and Pierre Curie shared a Nobel Prize in physics in 1903 with Henri Becquerel, who also studied radioactive materials.

Pitchblende is an ore of uranium. While working with pitchblende, Marie discovered that one of its elements, thorium, was also radioactive. In the same experiment, the Curies also discovered two new radioactive elements. Marie named these elements *polonium* (after her homeland, Poland) and *radium*.

In 1906, Pierre was run down and killed by a horse-drawn cart. Marie, who had once been refused a teaching job at the Sorbonne, took over Pierre's teaching duties. She became the first woman professor at the university. She never stopped her research on radioactive elements.

In 1911, years after its discovery, Marie Curie won a second Nobel Prize for her discovery of radium. Unlike the first prize, this Nobel Prize was in chemistry and was unshared. Although Pierre Curie was involved in the discovery of

PHYSICIST

radium, he could not share the award because he was dead. Nobel Prizes are not awarded posthumously. To date, Marie Curie is the only person to win Nobel Prizes in two different areas.

After winning her second Nobel Prize, poor health plagued Curie; however, it did not deter her from creating the Radium Institute, (a laboratory for the study of radiation) or from serving her country during World War I. During the war, Marie Curie trained people how to use X-rays to diagnose injuries. Marie herself set up X-ray stations throughout France and Belgium.

Ironically, the Curies never recognized the health hazards of working with radioactive materials. Many of those who worked in the lab died of anemia or leukemia. Marie Curie died of leukemia at age sixty-seven. Years later, the effects of the radiation also affected their daughter Irene-Joliot Curie, who was a scientist. Even today, the laboratory in which the Curies worked and their notes are still radioactive.

BUILDING ON THE PAST

Since Marie and Pierre Curie described radioactivity, many significant discoveries and achievements involving radioactive materials have occurred.

1903 Marie Curie shares Nobel Prize in physics with her husband Pierre Curie and Henri Becquerel for their discovery of radioactivity.

1911 Marie Curie wins Nobel Prize in chemistry for discovering the element radium. Curie is the only person to be awarded Nobel Prizes in two categories.

1917 Lise Meitner and Otto Hahn discover the element protactinium.

1935 Irene Joliet-Curie shares Nobel Prize in physics with her husband Frederic Joliet for creating artificial radiation.

1939 Lise Meitner and Otto Frisch publish a paper announcing that nuclear fission has been achieved.

1942 Enrico Fermi achieves the first sustained nuclear chain reaction. The event leads to the development of nuclear weapons and nuclear power.

1963 The first commercial nuclear reactor is opened at the Oyster Creek facility of Jersey Central Power.

1977 Rosalyn Sussman Yalow shares a Noble Prize in physiology or medicine with Soloman Berson for developing radioimmunoassay (RIA) tests. The tests use radioisotopes to measure hormones, enzymes, and proteins in blood.

1991 Physicists in England achieve controlled nuclear fusion.

Multicultural Women of Science ◆ Marie Curie

It's YOUR TURN

Hands-On Activity

CALCULATING THE HALF-LIFE OF URANIUM

MATERIALS (per individual)

graph grid, pencil, calculator (optional)

BACKGROUND INFORMATION

Radioactive elements decay, or break down, naturally. When uranium decays, it gives off enough particles to eventually change to lead. This decay takes place at a constant rate, called the *half-life*. The half-life of an element is the amount of time it takes for half the mass of a sample of the element to decay into another element.

The half-life of uranium is 4.5 billion years. So, in a ten-gram sample of uranium, only five grams of the sample will still be uranium in 4.5 billion years (one half-life). The other five grams (half) of the sample will have changed to lead. After two half-lives (9 billion years) only 2.5 grams of the sample is uranium (half the five grams present after the first half-life).

PROCEDURE

1. Make a table like the one shown below on a blank sheet of paper. Add six more rows to your table. Number the rows 1 through 6 in the first column.

Half-life	Mass of Uranium	Mass of Lead	Time Elapsed
Original Sample			0 years

2. On your graph grid (p. 162), outline a square 4 squares across by 4 squares down. Each square in the grid represents one gram of uranium. Count the number of squares in your grid and record this number under the head *Mass of Uranium* in the row titled *Original Sample*. Record *0* under the *Mass of Lead* heading.

3. Shade one-half of the squares in the grid. The unshaded squares are still uranium. The shaded squares show the mass of the uranium sample that changed to lead. Count the numbers of shaded and unshaded squares and record the data in the table. Record 4.5 billion years beneath the *Time Elapsed* head.

4. Shade one-half of the unshaded squares. Count the total number of shaded and unshaded squares and record the data. Add 4.5-billion years to the previous number in the *Time Elapsed* column.

5. Repeat step 4 four more times. Each time, record your data.

ANALYZE AND CONCLUDE

On separate paper, write your answers.

1. How much did the amount of uranium in the sample change from one half-life to the next?

2. Using the data in your table, how old would a rock be if it had equal amounts of uranium and lead?

3. After 28 billion years, is any of the sample still radioactive? If so, how much?

think WORK ACT

CRITICAL THINKING Answer the following questions in complete sentences.

1. What is radioactivity?

   _____

   _____

2. Describe three accomplishments of Marie Curie that were firsts for women.

   _____

   _____

3. Explain how radioactivity can be both helpful and harmful.

   _____

   _____

GOING FURTHER Complete three of the following.

BUILD YOUR PORTFOLIO

Use a periodic table to find the atomic number and atomic mass of radium. Use this information to draw a nucleus of a radium atom. Label the protons plus (+) and the neutrons zero (0).

ALTERNATIVE ASSESSMENT

Carbon-14 is a radioactive form of carbon that decays to form nitrogen. The half-life of carbon-14 is 5800 years. Design a way to answer the following: If a piece of bone contains 24 grams of carbon-14 when an animal dies, how much carbon-14 will the bone contain after three half-lives? How much of the original carbon-14 will have changed to nitrogen? How old will the bone be when it contains only 1.5g of carbon-14?

RESEARCH AND REPORT

Find out how radioactive elements are used to find the ages of rocks and fossils. Write a brief report of your findings.

JOURNAL WRITING

Marie Curie named one of the elements she discovered after her homeland. If you discovered a new element, what would you name it? Why?

IRÉNE JOLIOT-CURIE

1897-1956

Iréne Curie hated war. During World War I, she had seen much bloodshed and suffering while accompanying her mother, Marie Curie, to the front lines. Iréne assisted her mother in setting up X-ray units at field surgical hospitals and training doctors how to use them. Later, Iréne Curie defined science as, "the foundation of all progress that improves human life and diminishes suffering." Ironically, the work she did led to the development of a devastating weapon of war, the atomic bomb.

Iréne Curie was born in Paris, France, in 1897. She was a very bright and demanding child. When she was nine, tragedy struck the home. Her father, Pierre Curie, was killed in an accident. Her mother became a single parent with a full-time job. Fortunately, Pierre's father, who lived with the Curies, was there to help raise Iréne and her younger sister, Eve.

Iréne Curie had an unusual early education. Her mother and some of her colleagues at the Sorbonne did not approve of the schooling in Paris. They set up a cooperative school where they taught classes for a group of children that

Vocabulary

X-ray is a device that uses electromagnetic radiation to view the inside of an object or the human body.

included Iréne. The school closed after a short time and Iréne was sent to private school in 1910.

Iréne completed secondary school in 1914. She then went to the Sorbonne as a physics and mathematics student. Iréne earned both her B.A. and her M.A. in only two years. During this time, she also assisted her mother at the Radium Institute. It was at the Radium Institute that Irene became excited about research in radioactivity.

In December of 1924, Frederic Joliot began work at the Radium Institute. Joliot was a young engineer with little experience with research in radioactivity. Iréne was asked to teach Frederic the laboratory techniques developed by her parents. Less than two years after they began working together, Iréne and Frederic married.

The Joliot-Curies, who together had degrees in both chemistry and physics, made a good team. Their work led to many new discoveries by themselves and others. For example, James Chadwick, who applied experimental work done by the Joliot-Curies, was awarded a Nobel Prize in 1932 for his discovery of the neutron. C.D. Anderson, who was awarded a Nobel Prize in physics for his discovery of the positron in 1936, also based his studies on work done by the Joliot-Curies. In 1935, Frederic and Iréne received their own Nobel Prize. The prize was awarded in chemistry for their production of an artificial radioactive element from an element that is naturally stable.

As happened with Marie Curie, illnesses resulting from contact with radiation brought about the deaths of Frederic Joliot-Curie and Irene Joliot-Curie. In 1956, Irene died of leukemia, the same illness which had killed her mother. Two years later, Frederic died from liver problems caused by overexposure to radiation. During their life together, the Joliot-Curies raised two children, Helene and Pierre. In the family tradition, their daughter Helene also became a nuclear physicist.

BUILDING ON THE PAST

Many early scientists who studied radioactive materials developed serious illnesses or died as a result of their exposure to radiation. As scientists learned more about the dangers of exposure to radiation, new techniques and procedures were developed to make the use of radioactive materials safer. For example, lead shields are now used when X-rays are taken to protect people from excess exposure to radiation. In addition, people working near radioactive materials are required to wear badges that measure radiation exposure. Wastes from nuclear power plants are now encased in water and steel and then buried deep in the ground within a concrete shelter. All of these precautions are taken to prevent illnesses and diseases such as those developed by members of the Curie family.

It's YOUR TURN

Irene JOLIOT-CURIE

Hands-On Activity

CALCULATING HOW RADIOACTIVITY CHANGES WITH DISTANCE

MATERIALS

(per individual) metric ruler, calculator (optional)

BACKGROUND INFORMATION:

Radioactive materials give off particles and energy (radiation). This radiation, which can be measured, varies with distance from the source of radiation. The inverse square law illustrates this change. Steps for using the inverse square law are explained in the table.

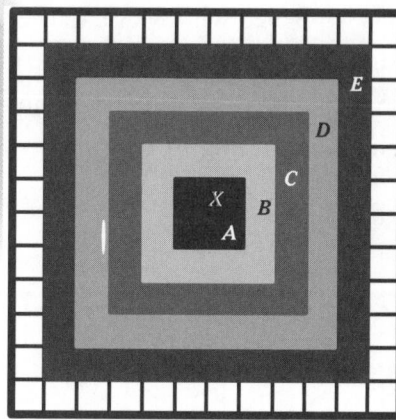

MOVING TOWARD THE SOURCE OF RADIATION	MOVING AWAY FROM THE SOURCE OF RADIATION
Calculate the change in distance between an object and the radioactive source.	Calculate the change in distance between an object and the radioactive source.
Write the change as a fraction e.g. 1/2	Write the change as a multiple of a number e.g. 2 × as far, 5 × as far
Invert the fraction by writing the bottom number as the top number and the top number as the bottom. e.g. 1/2 becomes 2/1 or 2	Square the number (multiply the number by itself) recorded in the previous step e.g. 2 × 2 = 4 or 5 × 5 = 25
Square the number (multiply the number by itself) recorded in the previous step. e.g. 2 x 2 = 4	Write the squared number from the previous step as a fraction. e.g. 4 becomes 4/1 or 25 becomes 25/1
The new measurement will be four times greater than the the previous measurement.	Invert the new measurement. e.g. 4/1 becomes 1/4 The new reading will be 1/4 of the first reading.

PROCEDURE

1. Look at the figure. The source of radiation is shown by *x*. The lettered regions show areas surounding the radiation.

2. Measure and record the distance from the radiation source to the outer border of area **B**.
3. Repeat step 2 for area **A**.
4. Record the change in distance moving from **B** to **A** as a fraction. Use the data in the table to calculate how much greater radiation is at A than at **B**.
5. Measure and record the distances from the source to area **E** and area **A**.
6. How much greater is the distance to **E** than to **A**?
7. How much less is a radiation reading at **E** than at **A**?

ANALYZE AND CONCLUDE

On separate paper, write your answers.

1. How does the amount of radiation change if the distance between an object and the source decreases by 1/4?
2. How does the amount of radiation change if the distance between the source and an object doubles?
3. If 6,000 units of radiation are measured at **A**, what will be the reading at **B**?

think WORK ACT

CRITICAL THINKING Answer the following questions in complete sentences.

1. What were some negative effects of the work done by the Joliot-Curies and other scientists studying radioactivity?

_____

_____

2. Do you think Irene Joliot-Curie would have continued her research if she knew it would someday be used to create an atomic bomb?

_____

_____

3. How do you think the work done by the Joliot-Curies may have differed if they had known about the dangers of working with radioactive materials?

_____

_____

GOING FURTHER Complete three of the following.

BUILD YOUR PORTFOLIO

A radioactive source gives off 10,000 units of radiation. Make a graph to show how the amount of radiation changes at five locations moving away from the source.

PERFECT YOUR SKILL

Elements with atomic numbers above 92 are radioactive. Make a table listing these elements, their atomic numbers, and the number of protons and neutrons for each. Use a periodic table for help.

RESEARCH AND REPORT

Both Marie Curie and Irene Joliot-Curie died of leukemia. Do research to find out what leukemia is and how it is treated. Prepare a report of your findings.

COOPERATIVE LEARNING

Radon is a radioactive gas present in the basements of many homes. Find out how radon gets into homes and how it is detected. Research the materials and methods used to test for radon. Conduct your own tests in three buildings. Report your findings to the class.

JOURNAL WRITING

The work of the Joliot-Curies contributed to the making of atomic bombs. In your journal, explain whether you think such bombs should be used during wartime. Give reasons for your response.

MEITNER
1878-1968

The nuclear age began around 1945. What people then only dreamed of doing with the energy of splitting atoms is today a reality. Many countries now use nuclear reactors to produce electricity or to power large seagoing vessels. Today's uses of nuclear energy resulted from years of research carried out by many scientists. Among them was Lise Meitner.

Lise Meitner was born in Austria in 1878. Her father Philip was a lawyer. Her mother Hedwick was a homemaker. Lise was one of seven children in a Jewish family. When Lise was young, formal education for women in Austria ended at age fourteen. When she turned fourteen, Lise began working as a governess, but was determined to continue her education on her own.

With a tutor and her own hard work, Meitner passed the entrance exam to the University of Vienna in 1901. In her first year, Lise became fascinated by the work done by the Curies. She decided to focus her studies on radioactivity.

At age twenty-seven, Meitner became only the second woman to earn a doctorate in physics from the University of Vienna. Two years later, in 1907, Meitner moved to Berlin. There she studied nuclear physics with Nobel

Prize winner Max Planck, who was working at the University of Berlin.

While in Berlin, Meitner also met Otto Hahn. Hahn was doing experiments on radioactivity at the Kaiser Institute of Chemistry. Hahn needed a physicist to work with him on his experiments. At the time, women were not allowed in the Institute; however, Hahn made arrangements for Meitner to set up a laboratory in the Institute's basement. To do this, Meitner had to agree not to enter any laboratories where men were working.

In a short time, Meitner's ability as a scientist was recognized. In 1918, Meitner and Hahn shared the directorship of the Institute. Meitner also became head of the Physics Department.

Working together, Meitner and Hahn discovered a new element, called *protactinium*. Later that same year, Meitner became the first woman physics professor at the University of Berlin.

> **Vocabulary**
>
> A **nuclear reactor** is a device in which nuclear fission occurs to produce energy for general use.

Meitner remained in Berlin for the next thirty-three years. She was forced to leave her home during World War II. Because of her Jewish faith, the Nazi threat in Germany placed her life in danger. Meitner fled to Sweden.

Throughout Meitner's career, much of her work focused on trying to make radioactive elements. In Sweden, Meitner continued her experiments with radioactivity. She worked with her nephew Otto Frisch. Hahn remained in Berlin. Hahn and Meitner often discussed the results of their experiments.

NUCLEAR PHYSICIST

To make radioactive elements, Hahn and Meitner added neutrons to uranium atoms. They did this by shooting neutrons at the nucleus. In December of 1938, Hahn made a startling discovery. When the uranium nucleus was bombarded with neutrons, it did not hold the particles. Instead, the nucleus split to form two lighter elements.

One month later, Meitner wrote an article about this discovery. She called this process of splitting the nucleus *fission*. During nuclear fission, great amounts of energy are released. Scientists immediately researched ways to use this energy. Ironically, the results of this research brought about the end to World War II, the war that had driven Meitner from her homeland.

In 1944, Hahn won the Nobel Prize in chemistry for the nuclear fission of uranium; however, it was Meitner who began the experiment and explained the process. Many scientists felt Meitner should have been included in the Nobel Prize awarded to Hahn. One year later, Meitner did get some recognition for her work. She was elected as a foreign member to the Swedish Academy of Science. This honor had been given to only two other women. About twenty years later, in 1966, Meitner and Hahn shared the Fermi Award given by the Atomic Energy Commission of the United States for their work on nuclear fission.

Meitner died in Cambridge, England, in 1968. Her contributions to nuclear physics will never be forgotten. In 1972, a new element (atomic number 109) was named *meitnerium* in honor of Lise Meitner.

APPLICATION OF THE SCIENCE

When the nucleus of an atom is split during fission, huge amounts of energy are given off. Scientists have found several ways to use this energy. One of the earliest uses of nuclear fission was in the development of the atom bomb, which was dropped twice in Japan in 1945, bringing an end to World War II. Peaceful uses of energy from nuclear fission were later developed. For example, nuclear reactors use fission to make electricity. In addition, nuclear energy also provides the power for ocean-going vessels, such as ice breakers, used by the United States and Russia.

It's YOUR TURN

Lise
MEITNER

Hands-On Activity
MODELING NUCLEAR FISSION

MATERIALS

(per group of two)

92 black beads, 144 white beads, shoe box lid

SAFETY

To prevent slips and falls, pick up any beads that fall onto the floor immediately.

BACKGROUND INFORMATION

In nuclear fission, an atomic nucleus is broken apart when it is struck by a neutron. This process forms new elements and may also give off neutrons. In the case of uranium, the new elements formed are barium and krypton as shown in the diagram. Barium has 56 protons and 85 neutrons. Krypton has 36 protons and 56 neutrons.

PROCEDURE

1. From your package of beads, remove one white bead. Set this aside.
2. Place all the other beads in the center of the cardboard. These beads represent the nucleus of a uranium-235 atom, which is made up of 92 protons (black beads) and 143 neutrons (white beads).
3. Reread the background information carefully. Using the black and white beads, make a model to show what happens when a uranium atom is split during nuclear fission. Place any extra beads off to the side.

ANALYZE AND CONCLUDE

On separate paper, write your answers.
1. What did the white bead you set aside in step 1 represent?
2. What was the total number of black and white beads used in your uranium model? In your krypton and barium models?
3. Describe the number and color of any beads that were not used in your model. What do these beads represent?
4. Predict what will happen if the extra "beads" continue moving and strike the nucleus of one or more neighboring uranium atoms.

NUCLEAR FISSION

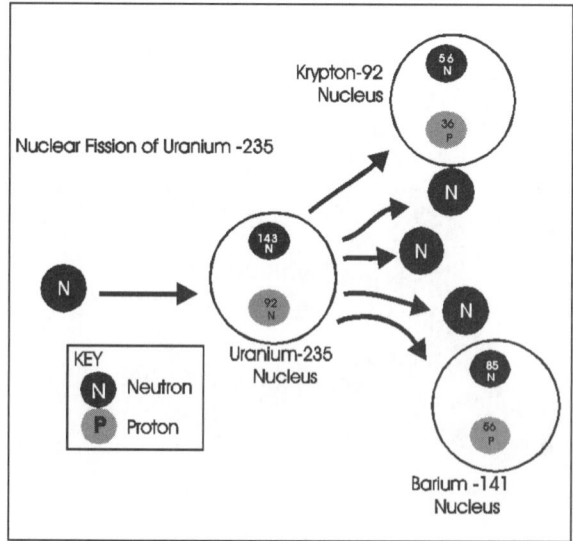

Figure 10

think WORK ACT

CRITICAL THINKING Answer the following questions in complete sentences.

1. Why did bombarding the nucleus of uranium with neutrons produce lighter elements instead of heavier elements?

   _____

   _____

2. In biology, the term *fission* describes a type of reproduction in which a cell and its nucleus divide to form two identical cells. How is this process similar to nuclear fission? How do the processes differ?

   _____

   _____

3. What are some ways that discrimination interfered with the work of Lise Meitner?

   _____

   _____

GOING FURTHER Complete three of the following.

BUILD YOUR PORTFOLIO

Make a drawing to show the elements and particles that result from the fission of a plutonium atom having 94 protons and 145 neutrons. Use the information in the Alternative Assessment to make your drawing. Be sure your drawing shows what happens to all neutrons.

ALTERNATIVE ASSESSMENT

Plutonium atoms with 94 protons and 145 neutrons may be used in nuclear fission reactions. This fission forms isotopes of the elements barium and strontium. The barium isotope is Ba-85. The strontium istotope is Sr-96. Use beads to show the atomic nuclei and particles formed in this fission reaction.

COOPERATIVE LEARNING

A chain reaction occurs when the neutrons given off by nuclear fission strike the nuclei of other radioactive atoms. Using the beads from the activity, work with two other groups to model a chain reaction that starts with uranium.

RESEARCH AND REPORT

Find out how nuclear fusion differs from nuclear fission. Write a summary of your findings. Include a diagram to explain the process.

Maria Goeppert-Mayer
1906-1972

In the 1920s, many scientists became excited about a new branch of physics called *quantum mechanics.* This science describes the behavior of atomic particles. Many early students of quantum mechanics attended the University of Göttingen in Germany. Among them was Maria Goeppert.

Maria Goeppert was born in 1906 in Kattowitz, Upper Silesia, then part of Germany, now located in Poland. Her father taught pediatrics at the University of Göttingen. He was a sixth generation university professor. Her mother, a former teacher, hosted many parties for prominent people. Maria enjoyed the social events hosted by her mother, but her real goal was to continue the family tradition of teaching at a university.

Maria enjoyed nature walks and fossil hunts with her father. Together, they observed eclipses and the motions of the moon. From these experiences, Maria developed an early interest in science. She studied hard, looking forward to the day she could enter the university.

Maria's curiosity and hard work paid off. She was accepted to the University of Göttingen. While there, she met such future Nobel Prize winners as Max Born, James Franck, and Enrico Fermi.

Maria also met Joseph Mayer, a chemist whom she eventually married. Maria and Joseph moved to Baltimore, Maryland, when Joseph took a teaching job at Johns Hopkins University. Despite her training, Maria could not get a university teaching job. Such jobs were not available to married women. Maria also went unrecognized for her research at the university, which she had to do as a volunteer.

Maria Goeppert-Mayer got her first paid teaching position at Sarah Lawrence College in 1941. She later worked with other noted physicists to develop the atomic bomb. Years later, in 1959, both she and Joseph accepted teaching positions at the University of California, San Diego.

Soon after moving to California, Goeppert-Mayer suffered a stroke. The stroke impaired her use of an arm and hand, but she never stopped working. In 1963, sixty years after Marie Curie shared the Nobel Prize in physics, Goeppert-Mayer became only the second woman to win this prize. She shared her award with J. Hans Jenson and Eugene P. Wigner. The three scientists won the prize for their description of the structure of the nucleus of the atom.

The description states that the protons and neutrons inside the nucleus of an atom are arranged in shells similar to the shells of the electrons that orbit the nucleus. The arrange-

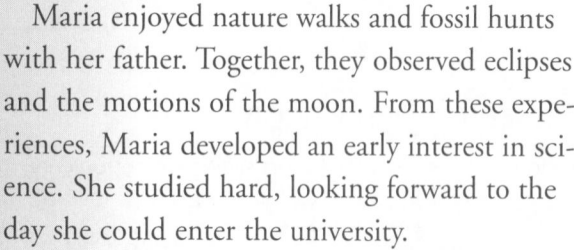

Vocabulary
The **nucleus** is the core or center of an atom

ment (coupling) and movement (orbiting) of protons and neutrons within their shells determines how stable an atom's nucleus will be.

Goeppert-Mayer continued her research and teaching after winning the Nobel Prize. Throughout her life, she received many honors and awards. She was also inducted into the National Academy of Sciences, the American Physical Society, the American Academy of Arts and Sciences, and the Academy of Heidelberg. At age sixty-five, Goeppert-Mayer died of heart failure.

BUILDING ON THE PAST

Developing a complete model of the atom involved the work of many scientists over several centuries. Each of these scientists built upon the ideas of other scientists from around the world. The major contributions of some of these scientists are summarized in the timeline.

TIMELINE

400 B.C.E. The Greek philosopher, Democritus, proposes that matter is made up of tiny, indivisible particles that are in constant motion. He uses the term *atom* to describe these particles.

1803 English chemist, John Dalton, states that all atoms of the same element are exactly the same and cannot be created, divided, or destroyed. He assigns atomic "weights" to the elements.

1897 English physicist, Joseph J. Thompson, discovers and identifies the electron.

1911 New Zealand physicist, Ernest Rutherford, describes the structure of the atom as a positively charged nucleus that is orbited by negatively charged electrons.

1913 Danish physicist, Niels Bohr, uses quantum theory to explain the electron orbit struc-ture of the atom.

1932 English physicist, James Chadwick, discovers and identifies the neutron.

1950 Maria Goeppert-Mayer publishes the shell and spin-orbit theory description of the nucleus of the atom.

It's YOUR TURN

Maria
GOEPPERT-MAYER

Hands-On Activity

DETERMINING NUCLEAR STABILITY OF ISOTOPES

MATERIALS

periodic table, pencil, calculator (optional)

BACKGROUND INFORMATION

Atoms of the same element all have the same number of protons in their nuclei, but the number of neutrons can vary. Atoms of the same element having different numbers of neutrons are called *isotopes.*

About 1,500 isotopes are known. Of these, only 264 are stable. The stability of the nucleus of an atom depends largely upon its neutron-to-proton ratio. Atoms having atomic numbers up to 20 are most stable if the ratio is one to one (one neutron for every proton). Those with higher atomic numbers should have a ratio closer to 1.5 to 1 (1.5 neutrons for every proton).

PROCEDURE

1. Use a periodic table to fill in the atomic number (Number of Protons) for each element in the table below.

2. Complete the *Number of Neutrons in Isotope* column by subtracting the atomic number of each element from its isotope mass. You may use a calculator.

3. Calculate the neutron-to-proton ratio for each isotope. To do this, divide the number of protons by the number of neutrons. Record the number you get, followed by a colon and the number 1 in the *Neutron-to-Proton* column. (e.g., 1.27:1)

4. Use the background information and your data to identify each element as stable or unstable. Record your data in the table.

ANALYZE AND CONCLUDE

On separate paper, write your answers.

1. What ratio of neutrons-to-protons did you get for stable isotopes with atomic numbers below 20? For stable isotopes with atomic numbers above 20?

2. How could you determine if the lead isotope Pb-206 is stable or unstable?

3. What does it mean when the nucleus of an atom is unstable?

Isotope	Isotope Mass	Element	Atomic Mass	Number of Protons/Atomic Number	Number of Neutrons in Isotope	Neutron to Protron Ratio	Stable or Unstable
C-12	12	Carbon	12				
Cu-66	66	Copper	64				
O-16	16	Oxygen	16				
N-16	16	Nitrogen	14				
Mn-56	56	Manganese	55				
S-32	32	Sulfur	32				
B-13	32	Sulfur	32				
Ca-42	42	Calcium	40				
Sn-118	118	Tin	119				
I-127	127	Iodine	127				

Multicultural Women of Science ◆ Maria Goeppert-Mayer   *© 1996 The Peoples Publishing Group, Inc.*

think WORK ACT

CRITICAL THINKING Answer the following questions in complete sentences.

1. Why was Goeppert-Mayer's description of the shell structure of the nucleus important?

   _____

   _____

2. Think about the work done by the Curies and the Joliot-Curies. How is the concept of a "stable" atom related to this work?

   _____

   _____

3. Describe some of the obstacles faced by Maria Goeppert-Mayer during her career.

   _____

   _____

GOING FURTHER
Complete three of the following.

BUILD YOUR PORTFOLIO

Find the number of protons and neutrons for elements 1 through 20 on the periodic table. Make a graph of your data. Use your graph to explain whether or not each element is stable.

PERFECT YOUR SKILL

Elements with atomic numbers of 92 and greater are radioactive, and therefore unstable. Use your knowledge of proton-to-neutron ratios to explain why these elements are unstable.

ALTERNATIVE ASSESSMENT

Make a table like that shown on p.116 that has five rows. Use a periodic table to fill in the table for the following elements: helium, uranium, lead, tungsten, and polonium. Which elements did you identify as stable? Why?

COOPERATIVE LEARNING

Hydrogen (atomic number 1) has three common isotopes. The most common isotope has 0 neutrons. Deuterium has 1 neutron, and tritium has 2 neutrons. In your group, make models to show the nuclei of these three isotopes. Label each model.

RESEARCH AND REPORT

Scientists have discovered that atoms are made up of subatomic particles in addition to protons, neutrons, and electrons. Do research to find the names of some of these particles and when they were discovered. Organize your findings in a timeline.

DOROTHY HODGKIN
1910-1994

Antibiotics are substances that kill or stop the growth of disease-causing bacteria. Penicillin was the first antibiotic. Today, many antibiotics exist. Antibiotics are often made naturally by molds and bacteria. They may also be made synthetically. Much of the technology used to make synthetic antibiotics resulted from the work of Dorothy Crowfoot Hodgkin.

Dorothy Crowfoot was born in Cairo, Egypt, in 1910. Her father, John Winter Crowfoot, was an archaeologist working in Egypt. At the time, Egypt was a British colony. With the threat of war in the region, Dorothy and her sisters were sent to England, where they spent most of their youth.

After the war, Dorothy visited her parents in Africa during summers. When she was thirteen, Dorothy met A. F. Joseph, a soil chemist and family friend. This meeting led Dorothy to develop an interest in minerals, especially crystals and their structures. Crystals are solids that have a definite chemical makeup and shape. Their shape is due to the arrangement of their atoms. Crystals formed in living things are called biological

> ## Vocabulary
> Molecular structure is the exact atomic composition and arrangement of atoms amd molecule s.

crystals. Examples of biological crystals are the pepsin produced by the stomach and the insulin produced by the pancreas.

At age sixteen, Dorothy learned that X-rays could be used to identify the atomic structure of crystals. The use of X-rays on biological crystals became her life's work. This work began in 1928 when Dorothy entered Oxford University, specializing in X-ray crystallography. She completed her studies in 1932. Later that year, she went to Cambridge University to do X-ray analysis research.

By 1934, a married Dorothy Hodgkin was suffering from severe rheumatoid arthritis. No effective treatment for the disease existed; however, Hodgkin continued to work despite the pain and limited use of her hands.

Hodgkin spent five years figuring out the structure of penicillin. She spent another eight years finding the structure of vitamin B_{12}. At first, she had to use mechanical machines to make her mathematical calculations. Later, she used computers which carried out her calculations more quickly.

In the fall of 1964, Hodgkin was working in Africa. There she learned that she had won a Nobel Prize in chemistry. Hodgkin is only the third woman to win this prize. She was given the award for finding the molecular structure of vitamin B_{12}. When Hodgkin began her work, she was trying to figure out the crystal structure of insulin. She solved this mystery

CRYSTALLOGRAPHER

years after beginning her research. The work done by Hodgkin laid the foundation for the making of penicillin, vitamin B_{12}, and insulin in the laboratory. This, in turn, made it possible to make these substances widely available to people who needed them.

Hodgkin traveled extensively, sharing her scientific knowledge with others. She has many awards, including England's Order of Merit.

The only other woman to receive this award was Florence Nightingale. Hodgkin also lunched with British Prime Minister, Margaret Thatcher. Thatcher was once a chemistry student of Hodgkin. In July 1994, Dorothy Hodgkin died. The goals and techniques of her work live on in the branch of science called molecular biology.

APPLICATION OF THE SCIENCE

Before 1922, people diagnosed with diabetes usually died from the disease. Diabetes occurs when the pancreas does not produce enough insulin. Insulin is a hormone that allows the body to break down and use sugar.

In 1921, doctors John Macleod, Frederick Banting, and Charles Best found a way to remove insulin from the pancreas of dogs. About a year later, this insulin was successfully used to treat human patients. As demand for insulin taken from dogs became greater than the supply, larger animals such as oxen, sheep, and pigs had to be used.

Hodgkin determined the structure of insulin in 1969. Her research paved the way for the production of insulin in the laboratory. Today, human insulin is prepared using bacteria in a process called genetic engineering. This allows insulin to be made in large enough amounts to meet human demand.

1996 The Peoples Publishing Group, Inc. **Multicultural Women of Science** ◆ Dorothy Hodgkin

It's **YOUR TURN**

Dorothy
HODGKIN

Hands-On Activity

GROWING CRYSTALS

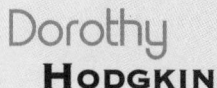

MATERIALS (per group of three)

3 clean 100-mL beakers, alum, sugar, salt, heat source, glass stirring rod, three pieces of uncoated string, 3 plastic-coated paper clips, 3 straws, hand lens, beaker tongs, wooden splint (spatula), glass marking pencil

SAFETY

Wear safety goggles and a laboratory coat or apron throughout this activity. Clean up any spills that occur immediately. Be careful using the heat source to avoid burns. Handle hot objects carefully.

BACKGROUND INFORMATION

A solution forms when one substance dissolves in another. In a solution, the substance that dissolves is called the solute. The substance in which the solute dissolves is the solvent.

When a solution cannot hold any more solute at a given temperature, it is said to be saturated. In most cases, heating a saturated solution increases its ability to hold more solute, causing the solution to become supersaturated. As a supersaturated solution cools, the excess solute comes out of solution and forms crystals.

dissolves. The excess alum should settle at the bottom of the beaker.

4. Gently heat the solution while stirring it. As the alum at the bottom dissolves, add more alum. Continue adding alum and stirring until no more alum dissolves.
5. Using the beaker tongs, remove the beaker from the heat. Set one of the straws across the top of the beaker so the string and clip are suspended in the solution.
6. Repeat steps 1 through 5 using first sugar, then salt.
7. Set the beakers where they will be undisturbed. Observe each beaker daily for one-to-two weeks. Record your observations daily.

PROCEDURE

1. Tie one end of a string around the center of a straw and the other end to a plastic-coated paper clip. Place the straw across the top of the beaker and lower the string into the beaker, making sure the clip does not touch the bottom of the beaker.
2. Remove the straw and string from the beaker and set it aside. Repeat steps 1 and 2 using the other straws, clips, and string.
3. Label a beaker with your group name and number and the word *alum*. Fill the beaker with 50 mL of water. Slowly add alum to the water, stirring the mixture until no more alum

ANALYZE AND CONCLUDE

On separate paper, write your answers.

1. At what stage in the procedure did you get a saturated solution? A supersaturated solution?
2. Why was the solution heated?
3. What caused the formation of crystals on the string?
4. List all the similarities and differences you observed in the formation of the different types of crystals?
5. What conditions are necessary for crystals to form?

120

think WORK ACT

CRITICAL THINKING Answer the following questions in complete sentences.

1. How was Dorothy Hodgkin able to determine the structure of the crystals that she grew?

   _____

   _____

2. What causes crystals to have a definite shape?

   _____

   _____

3. Explain how you could use a hand lens to distinguish between sugar and salt.

   _____

   _____

GOING FURTHER Complete three of the following.

BUILD YOUR PORTFOLIO

In your portfolio, make drawings to show several types of crystal shapes. Label each shape. Use library resources if necessary.

ALTERNATIVE ASSESSMENT

Do research to find the names and characteristics of the different crystal groups. Make models to represent each crystal group.

RESEARCH AND REPORT

Write to your state chamber of commerce or Dept. of Natural Resources to find out where in your state crystals form naturally. Combine the information you gather in a state map that can be shared with the class.

COOPERATIVE LEARNING

Do library research to find out about some uses of crystals other than in medicine. Combine your findings in a group table.

JOURNAL WRITING

In your journal, write a letter to Dorothy Hodgkin explaining why you think her work was important.

ROSALYN YALOW

1921-

Have you ever had a blood test? Today, blood tests tell much about individuals. For example, blood tests can identify certain diseases and show the level of glucose (sugar)or alcohol in the body. Athletes are given blood tests to be sure they are not using steroids or other drugs. Many uses of blood tests are possible because of the work of Rosalyn Sussman Yalow and her partner Solomon A. Berson.

Rosalyn Sussman was born in the Bronx, New York, in 1921. As a young girl, she decided she would someday have both a career and a family. She began working on the first part of her goal as a physics student at Hunter College.

Rosalyn Sussman graduated Hunter College, New York City, with honors in 1941. She then looked for a job as a teaching assistant at the university level. Sussman was turned down for positions at many universities because she was female and Jewish. It was not until the United States became involved in World War II and many male college professors left their jobs for the military that Rosalyn Sussman was able to get her first teaching position.

Sussman's first teaching position was at the University of Illinois, a school which had once turned her down. While there, she met and married Aaron Yalow, a graduate student. In 1945, she earned her Ph.D. in nuclear physics. Soon after, the couple set up house in New York and began to build their careers. They also began the second part of Rosalyn's goal—raising a family.

In 1947, Rosalyn Yalow began a new position at the Bronx Veterans Affairs Medical Center. Three years later, she met and began working with Dr. Solomon A. Berson, a resident in internal medicine. For twenty-two years, Yalow and Berson worked together and they developed a way to use radioactive elements to detect certain substances in body tissues. The procedure is called *radioimmunoassay* (RIA).

In RIA, a small amount of a radioisotope is placed in a tissue sample from the body. The radioisotope binds with certain proteins, allowing them to be detected and traced. Using RIA, the activities of some proteins, such as enzymes and hormones, can be studied. Using this information, RIA can be used to determine the rate at which substances are produced by or used in the body.

In 1973, Rosalyn Yalow was elected to the National Academy of Sciences. The following year, Rosalyn Yalow became the first woman to win the Albert Lasker Basic Medical Research

Vocabulary

Radioimmunoassay (RIA) uses radioactive elements to detect certain substances in body tissues.

NUCLEAR PHYSICIST

Award—the highest science award in the United States. In 1977, she became the second woman to win the Nobel Prize in physiology or medicine for the development of RIA. Yalow's partner, Dr. Solomon Berson, did not share in the prize because he died in 1972. Roger C. Guillemin and Andrew V. Schally shared the prize that year for their discovery of the part of the brain that controls endocrine functions.

Today, Yalow spends much her time speaking about the importance of a good education. She encourages young people, especially young women, to enter the sciences. Yalow also speaks about the need for good child care centers. She believes that such centers help women who work outside the home to reach their true potential.

APPLICATION OF THE SCIENCE

In RIA, radioisotopes are used as tracers or indicators to identify various chemical reactions within the body. Other indicators are used to identify acids and bases. The term *pH* followed by a number is used to describe how acidic or basic a substance is. Distilled water is neutral (neither acid nor base) and is assigned a pH of 7. Acidic substances have pH values less than 7. Basic substances have pH values greater than 7.

Indicators that identify pH change color in the presence of an acid, a base, or both an acid and a base. For example, red litmus paper turns blue in a base, but remains red in an acid or a neutral substance. Blue litmus paper turns red in an acid, but does not change color in a base or a neutral substance. Other indicators are used to identify the exact pH value of substances.

It's YOUR TURN

Rosalyn YALOW

Hands-On Activity

USING INDICATORS TO IDENTIFY ACIDS AND BASES

MATERIALS (per group of four)

red and blue litmus paper (ten pieces each), container of red cabbage juice, ten 1-oz. clear plastic cups, dropper, milk, vinegar, household ammonia, soap, lemon juice, club soda, antacid, distilled water, (two other household items such as mouthwash, toothpaste, sugar, salt, bleach, or dishwasher detergent.)

SAFETY

Wear safety goggles and a laboratory coat or apron throughout this activity. Strong acids and bases can burn your skin. If you get a substance on your skin, wash the area immediately with water. If a spill occurs, notify your teacher immediately so it can be cleaned up in the appropriate manner.

BACKGROUND INFORMATION

Many chemicals can be used to indicate the presence of an acid or a base. Such chemicals include red cabbage juice, bromthymol blue, phenolphthalein, and congo red. When using cabbage juice as an indicator, the cabbage juice remains a deep purple-blue color in a neutral substance. In a base, the cabbage juice turns light blue or green-blue. In an acid, the cabbage juice turns red or pink. The color change of some indicators identifies the exact pH of a substance. Other indicators provide only general information about whether a substance is acidic, basic, or neutral.

PROCEDURE

1. On a sheet of paper, make a table like the one shown. Add enough rows to record data for the ten items you will test.

Substance	Red Litmus Paper Color	Blue Litmus Paper Color	Color of Cabbage Juice Indicator	Acid or Base?
vinegar				

2. Label each cup with the names of the substances to be tested. Place a few drops of each substance into the proper cup.
3. Touch a piece of red litmus paper to each substance. Record in the table the color of the litmus paper after you touch the substance.
4. Repeat step 3 using the blue litmus paper. Record your data.

5. Use the dropper to add 2-3 drops of red cabbage juice to the ammonia. Observe the resulting color and record it in your table.
6. Rinse the dropper thoroughly with water.
7. Repeat steps 5 and 6 (one substance at a time) for each of the remaining substances.
8. Dispose of all substances as instructed by your teacher.

ANALYZE AND CONCLUDE

On separate paper, write your answers.

1. What substances, if any, did you identify as neutral? Why?
2. What substances did you identify as acids? Why?
3. What substances did you identify as bases? Why?
4. Why was it necessary to wash the dropper between each test?

think WORK ACT

CRITICAL THINKING Answer the following questions in complete sentences.

1. Radioactive isotopes can be detected using a Geiger counter. How might a Geiger counter and radioactive isotopes be used to detect water main leaks?

_____

_____

2. How is the use of pH indicators similar to that of radioisotopes?

_____

_____

3. Steroids and other drugs can be detected in the blood through the use of indicators. Why might such tests be given to athletes? Do you think this is a good practice? Explain.

_____

_____

GOING FURTHER Complete three of the following.

BUILD YOUR PORTFOLIO

If your school has a sports program, survey the coaches and student athletes to find out their feelings about drug testing of athletes. Prepare a bar graph of your data.

ALTERNATIVE ASSESSMENT

Obtain pH paper from your teacher. Use the pH paper to test each of the substances you used in the activity. Record the pH of each substance and compare the values you obtain with the results you obtained in the activity.

JOURNAL WRITING

Many employers require drug testing of job applicants. In your journal, explain whether you think such testing is necessary or is a violation of an applicant's rights.

COOPERATIVE LEARNING

Conduct a survey of cosmetics and hair care products in a local pharmacy. Look for items that contain information about pH on the label. Record the product name and identify the pH value or pH information given on the label. As a group, discuss the items you observed. Make a conclusion about why this information is included on the label.

RESEARCH AND REPORT

Do library research to find out what specific diseases are diagnosed using RIA. Write a report of your findings.

WOMEN OF

DIAN FOSSEY
1932-1985

RACHEL CARSON
1907-1964

ENVIRONMENTAL SCIENCE

ANN HAVEN MORGEN
1882-1966

YNES MEXIA
1870-1938

RACHEL CARSON
1907-1964

The phrase, "The pen is mightier than the sword," first appeared in a historical drama published in 1839. Since that time, its interpretation has been a topic of much discussion. This phrase sums up the work of ecologist and marine biologist, Rachel Carson. In 1962, Carson published the book *Silent Spring*. The book focused attention and began a debate on the use and effects of pesticides. It is a debate that continues today.

Rachel Carson was born in the farming community of Springdale, Pennsylvania, in 1907. As a child growing up on a farm, Rachel was fascinated by the wildlife around her. Her fascination and respect for nature was encouraged by her mother.

Rachel was educated in the local public schools. After graduating from high school, she enrolled in the Pennsylvania College for Women. Her first career choice was writing; however, she changed her major from English to zoology after taking a biology course. Zoology is the branch of science that deals with animals.

In 1929, Carson received her bachelor's degree in zoology. That same year, she enrolled at Johns Hopkins University. She taught summer sessions there while working toward her master's degree. In 1931, Carson began teaching zoology at the University of Maryland. The following year, she received her master's degree from Johns Hopkins.

Carson resigned from the university in 1936 to become a marine biologist for the United States Bureau of Fisheries. The bureau later became the United States Fish and Wildlife Service. Carson was one of only two women working for the agency in a nonclerical position. She used her writing skills and expertise in science to write many of the informational booklets and leaflets issued by the agency. Eleven years later, she became the Editor in Chief for the Bureau.

Between 1941 and 1964, Rachel Carson published five books, each of which includes an unusual and enjoyable mix of science and literature. Carson's first book, *Under the Sea*, was published in 1941. The book described life at the shore, the open sea, and on the bottom of the sea. In 1951, her second book, *The Sea Around Us* was published. This book describes the history, physical features, chemistry, and biology of the sea. The book was an instant success. In less than six months, it went into its ninth printing, became a national best-seller, and later won the National Book Award in nonfiction. But the

Vocabulary

Biology is the study of all forms of life. This broad area of science is divided into specialized branches. For example, zoology deals with the study of animals, ecology with the study of the interaction between living things and their environment.

most notable and controversial of her books was *Silent Spring*, published in 1962. This book established Carson as an ecologist. *Silent Spring* would later come to have a huge impact on environmental legislation in the United States of America.

Silent Spring reflected much of Carson's work on ecology during her years at the Bureau. Carson was increasingly concerned with the widespread use and likely dangers of pesticides, especially DDT. She was concerned that this and other pesticides introduced into the environment cycled through the food chain. She argued that the effects of many pesticides had not been fully studied. Similarly, she did not accept the belief that environmental damage was a price that had to be paid for technological development.

Carson died in 1964, but her efforts for a safer environment continued long after her death. *Silent Spring* not only focused attention on the dangers of pesticides, but also brought a halt to the use of many of them. The President's Science Advisory Council strongly endorsed Rachel Carson's claims. As a result, several laws were passed that placed major restrictions on the use of dangerous pesticides.

IMPLICATION OF THE SCIENCE

Scientists have shown that insecticides, such as DDT, can collect in body tissues and increase in concentration as they pass through the food chain. This process is called *bioaccumulation*. Such a process led to a disruption in the formation of the calcium carbonate present in the shells of the eggs of bald eagles, perigrine falcons, and other predatory birds. Without the correct amount of calcium carbonate in the shells, the eggs cracked, preventing the embryos inside from developing. As fewer offspring were produced, a decline in the populations of these birds took place. The connection between the declining bird population and DDT led to the ban of DDT in 1972. Since the ban took effect, the population sizes of bald eagles and perigrine falcons have increased. In fact, the bald eagle population has increased so much, the species was taken off the endangered species list in the mid-1990s.

Rachel CARSON

It's YOUR TURN

Hands-On Activity

OBSERVING CHEMICAL CHANGE INVOLVING CALCIUM CARBONATE

MATERIALS (per group of two)

medium-sized wide-mouthed jar with lid, an egg, vinegar (about 100 mL)

SAFETY

Wear goggles and a laboratory coat or apron throughout this activity. Clean up any spills that occur immediately.

BACKGROUND INFORMATION

Egg shell and bone are rich in calcium minerals. Bone contains calcium phosphate and calcium carbonate. Egg shell is rich in calcium carbonate. The presence of calcium compounds in egg shell and bone is responsible for the hardness of these materials. Calcium dissolves easily in the presence of an acid, such as the hydrochloric acid in your stomach, and acetic acid, commonly called vinegar. As calcium compounds in egg shell and bones break down, these structures lose their hardness. They become rubbery or brittle and lose many of their important properties, including those of support and protection.

PROCEDURE

1. Observe an egg. Pay particular attention to the appearance and texture of its shell.
2. Carefully place the egg into the jar.
3. Add vinegar to the jar until the liquid level is about 2.5 cm (1 inch) above the top of the egg.
4. Tightly close the lid on the jar. Place the jar where it will remain undisturbed overnight.
5. Remove the egg from the jar. Examine the appearance of the egg. Pay particular attention to the appearance and texture of the shell. Record your observations.
6. Repeat steps 4 through 6 for one week.

ANALYZE AND CONCLUDE

On separate paper, write your answers.

1. What changes, if any, did you observe in the appearance of the shell of the egg? What do you think caused these changes?
2. Did you observe any other changes in the egg, besides the changes in the shell. If so, describe the changes.

think WORK ACT

CRITICAL THINKING   Answer the following questions in complete sentences.

1. Explain how the phrase "the pen is mightier than the sword" applies to the work of Rachel Carson.

   _____

   _____

2. Explain how Rachel Carson's predictions about the use of chemical pesticides proved to be correct years after her death.

   _____

   _____

3. How is the work of Rachel Carson still affecting the way people look at the environment?

   _____

   _____

GOING FURTHER   Complete three of the following.

BUILD YOUR PORTFOLIO

Survey your community for evidence of environmental problems, such as pollution. Describe in detail five such problems and the locations they affect or photograph the areas and describe the problems shown in captions.

ALTERNATIVE ASSESSMENT

Design an experiment to show how fertilizers or pesticides sprayed onto farmland or a field may enter the water supply. Submit your experimental design to your teacher for approval, comments, and suggestions.

JOURNAL WRITING

In your journal, describe things you do to help your environment. Explain the benefits of each activity you describe.

COOPERATIVE LEARNING

Organize a cleanup project for an area in your community or a section of your school grounds that needs attention. Prepare materials to alert people in your community about the problems you identify. Get volunteers to sign up for and carry out your cleanup activity

RESEARCH AND REPORT

Do research on bioaccumulation as it relates to DDT and the bald eagle population. Prepare an oral report of your findings. Include artwork that explains the problem in your presentation.

Environmental concern is focused on the destruction of the world's rain forests. Such destruction will bring about the extinction of many plant and animal species. Many rain forest plants contain chemicals that are used to treat illnesses. Medicines now made from rain forest plants include taxol, a cancer fighting drug, and quinine, which is used to treat malaria. Many other medicines may be hidden in rain forest plants; however, if the forests are destroyed, the medicines held in these plants will go undiscovered. Ynes Mexia, a Mexican American botanist and explorer, was an early pioneer in this field.

Ynes Mexia was born in Georgetown, Washington, D.C. Her father, Enrique Mexia, worked for the Mexican government. Her mother, Sarah Wilmer, raised Ynes and six children from a previous marriage. Ynes had a troubled childhood marked by frequent moves. At age three, she moved to Texas, where she lived until she was fifteen. Little is known of Ynes's early education; however, in 1886, Ynes moved again, first to Philadelphia and then to Maryland.

While in Maryland, Ynes attended St. Joseph's Academy in Emmitsburg until 1887.

Vocabulary

Genus is a group of several species with many similarities.

She then went to Mexico City, where she lived for ten years in her father's home. In 1897, at age twenty-seven, Ynes married Herman Lane, a Spanish-German merchant. Ynes's marriage ended with the tragic death of her young husband in 1904. In 1907, she married Augustin A. de Reygados. This marriage ended in divorce.

Mexia moved to San Francisco, California, in 1908. Her divorce left her depressed and withdrawn. She kept to herself for long periods of time. In 1920, Mexia became interested in flowers, often taking field trips with the local Sierra Club. In 1921, at age fifty-one, she was admitted to the University of California at Berkeley. She became interested in the natural sciences, especially botany. At a time when most people begin thinking about retirement, Ynes Mexia's life as a botanist, adventurer, and explorer, was about to begin.

Mexia entered her new life with enthusiasm. At age fifty-five, having completed three years of college study, Mexia made her first major excursion. She went to western Mexico where she collected hundreds of plant specimens. Mexia continued her botanical studies by exploring the remote slopes of Mount McKinley, in Alaska. This was followed by a trip to Brazil, and a thirty-month excursion into the forests of the Amazon. Mexia described her two-and a-half year Amazon adventure in her writing, *Three Thousand Miles Up the Amazon,* in the Sierra Club Bulletin.

Mexia traveled by ship, dugout canoe, and balsa raft, collecting and classifying plant specimens. She lived on sandy beaches under mosquito netting and bartered with local peoples for food.

Mexia returned to the United States in 1932. She was sixty-two years old and passionate about her life's work. She traveled to California, Nevada, Utah, and Arizona collecting, describing, and photographing plants. Next, she journeyed into the forests of Peru, Ecuador, Chile, and Argentina. In 1937, she again returned to the United States, where she took classes at the University of Michigan. The following year,

after returning from a trip to southwestern Mexico, Ynes Mexia died of lung cancer.

During the course of her work, Mexia collected more than 150,000 plant specimens. She discovered one new genus and more than 500 new species. Many of these plants were named in her honor. In addition to describing and classifying new plants, Mexia provided drawings that showed them in their natural habitats. Her accurate observations and descriptions validated and updated the work of earlier botanists. Today, her field notes are kept as a reference at the University of California at Berkeley.

APPLICATION OF THE SCIENCE

Ynes Mexia often stayed with Native Americans during her exploration into the remote areas of North, Central, and South America. In some remote rain forest areas, very small societies still exist. In these groups, healers, called shamans, tend to the sick. The shamans often use plants that grow in the area to treat illness. Knowledge of the healing powers of the plants is passed from one shaman to the next.

Some botanists study plants and the cultures of different peoples. These scientists often seek out shamans. By knowing the language and native ways of the shamans and their people, the scientists hope to learn about the healing powers of the plants used by the shamans. Plants that show potential as medicines are then sent to drug companies for testing. Botanists who study native peoples are hopeful that the knowledge they gain will help them find the medicines needed to cure many diseases in less time than it would take to find these plants on their own.

It's YOUR TURN

Ynes
MEXIA

Hands-On Activity

IDENTIFYING TREES BY THEIR LEAVES

MATERIALS (per student)

crayon, six sheets of white typing paper, notebook or clipboard, field guide to trees

PROCEDURE

1. Find a leaf that has fallen to the ground. Place the leaf smooth side down on a hard surface such as a notebook or clipboard.

2. Cover the leaf with a sheet of paper. Use the crayon to gently rub across the surface of the paper until a clear image of the leaf appears. Label the diagram *Figure 1*.

3. Examine the leaf. On the back of your tracing, list as many characteristics of the leaf as possible. Include information about color, vein pattern, and other information you observe about leaves of the same type that are still on the tree. Use the illustrations to classify your leaf as simple, compound (palmate) or compound (pinnate).

4. Repeat steps 1 through 3 for four other kinds of leaves. Label the drawings as *Figures 2* through *5*.

5. Locate a field guide. Match your tracing to a photograph in the guide. From your observations, identify the trees from which your leaves came. Write this information beside the Figure number of your tracing.

SAFETY

Wear appropriate clothing when working outdoors. Avoid contact with poisonous plants such as poison ivy and sumac.

ANALYZE AND CONCLUDE

On separate paper, write your answers.

1. How many of your leaves were you able to identify?

2. What characteristics helped you identify each leaf?

3. Compare your leaf tracings with those of other students. Based on this information, which types of trees seem to be most common in your area? Least common?

4. What information besides leaf traits would be helpful in identifying trees?

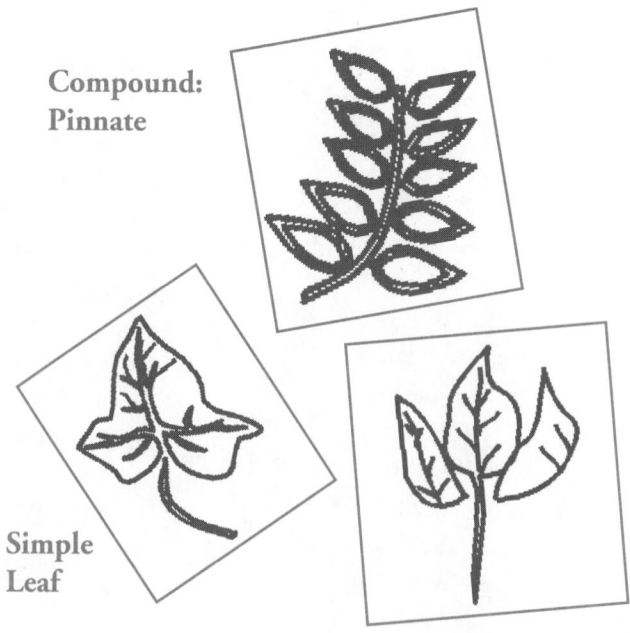

Compound: Pinnate

Simple Leaf

Compound: Palmate

Multicultural Women of Science ◆ Ynes Mexia

think WORK ACT

CRITICAL THINKING Answer the following questions in complete sentences.

1. Botanists are scientists who study plants. Why is the work of botanists important?

   _____

   _____

2. The work done by Ynes Mexia was done as field studies. What advantages do you think working in the field has over working in a laboratory?

   _____

   _____

3. How might the work done by Ynes Mexia be helpful to scientists studying tropical rain forests today?

   _____

   _____

G O I N G   F U R T H E R Complete three of the following.

BUILD YOUR PORTFOLIO

Create a pictorial or photo essay of plants common to your area. Include ten trees, ten grasses, and ten flowers. Label each plant with its common and scientific names.

JOURNAL WRITING

In your journal, describe which area visited by Ynes Mexia you would most like to explore. Give reasons for your choices. What would you most like to learn about the area?

PERFECT YOUR SKILL

Use information in the reading passage to develop a timeline that outlines the life and accomplishments of Ynes Mexia.

COOPERATIVE LEARNING

Work with three classmates to create a field guide to the plants in your area. Use library resources to find out about each plant. Include drawings or photographs in your guide. Also include information such as each plant's scientific name, its features, and its uses.

RESEARCH AND REPORT

Use library resources to find out about the drug taxol. Prepare a report explaining how taxol is used and where it comes from.

ANN HAVEN MORGAN
1882-1966

Earth is the only planet known to support life. Biology is the area of science that studies living things. This broad area is divided into many specialized branches. For example, zoology deals with the study of animals. Botany is the study of plants. Ecology is the study of the way living things interact with each other and with their environments. Many biologists focus their work in only one area; however, Anna (Ann) Haven Morgan is a scientist who used ideas of ecology and conservation in her work as a zoologist.

Anna Morgan was born in Waterford, Connecticut, in 1882. After graduating from secondary school in 1902, Morgan went to Wellesley College. She then transferred to Cornell University two years later. Morgan graduated from Cornell in 1906 and became a zoology instructor at Mount Holyoke College. She worked at Mount Holyoke for three years before returning to Cornell for graduate study.

While in graduate school, Morgan became interested in aquatic insects. In 1912, she received her Ph.D. from Cornell for her studies of mayflies. Her excitement about these insects earned her the nickname "Mayfly Morgan." That same year, she also changed her name from Anna to Ann.

> ### Vocabulary
> Ecosystem describes all the relationships and interactions that make up an organism's environment.

Morgan returned to her teaching position at Mount Holyoke after receiving her Ph.D. She introduced many new courses at the college, including one dealing with the biology of water organisms and another with animal life in winter. One of her books, *Field Book of Animals in Winter*, published in 1930, resulted from the courses she taught.

In 1916, Morgan became chairperson of the zoology department of Mount Holyoke. She held this position until her retirement in 1947. During summers, Morgan did field work at research laboratories. Between 1918 and 1923, she taught courses about sea stars and other similar marine animals at the Marine Biological Laboratories in Woods Hole, Massachusetts. In 1926, she did research at the Tropical Laboratory in Kartabo, British Guiana.

Morgan published many articles and books that included her drawings and photographs. Unlike many texts in science, Morgan's books were popular with the general public. They were interesting, scholarly, beautifully illustrated, and written in an easy-to-understand language. In addition to serving as field guides, the books included topics on conservation and ecology. In her *Field Book of Ponds and Streams: An Introduction to the Life of Fresh Water*, Morgan identified the traits as well as the habitat of each animal group. She also showed the relationship of each animal group to others and explained each group's place within the ecosystem.

ZOOLOGIST
ECOLOGIST

In 1955, Morgan published her last book, *Kinships of Animals and Man: A Textbook of Animal Biology*. This book encouraged people to preserve the natural environment. After publishing this book, Morgan spent much of her time trying to change the way science was taught in schools. She wanted topics about ecology to be included in science courses. To aid in this goal, Morgan became a member of the National Committee on Policies in Conservation Education. She also held workshops on ecology and conservation for teachers. This work earned Morgan a reputation as a pioneer in the ecological movement of the United States. In 1966, Morgan died of stomach cancer. She was eighty-four.

APPLICATION OF THE SCIENCE

In the early part of this century, science education began to emphasize nature studies. These studies focused on recognizing and classifying plants and animals. As education in these areas developed, a new direction began to emerge in science. This direction resulted from the work of Ann Haven Morgan and others who not only identified and classified plants and animals, but studied their relationship to their environment. Morgan sought to include these topics in high school and college science courses.

Today, the subjects of ecology and environmental science have become a major part of science education. As a result of these studies, people have become more aware of the need to respect and preserve the environment and its resources.

It's YOUR TURN

Ann Haven MORGAN

Hands-On Activity

MAKING AN AQUATIC ECOSYSTEM

MATERIALS (per group of four)

one gallon jar, gravel, water plants (such as *Elodea*, *Spirogyra*, or *Cabomba*), goldfish, snails, gooseneck lamp with sixty-watt bulb

SAFETY

Handle the glass jar carefully to avoid breakage. Treat all living things with care and respect. Clean up any spills immediately.

BACKGROUND INFORMATION

An ecosystem is all the living and nonliving parts of an environment and their interactions. Ecosystems provide organisms with all their needs. For example, air, proper temperature, food, and water are all parts of an organism's ecosystem.

PROCEDURE

1. Wash the jar thoroughly with water.
2. Add gravel to the bottom of the jar.
3. Gently insert the plants into the gravel. Fill the jar with water and allow it to sit undisturbed for one or two days. This allows the water temperature to reach room temperature and allows particles in the water to settle.
4. After two days, carefully add some goldfish and one or two snails to the water. Set up a gooseneck lamp with a 60-watt bulb about 18" from the jar.
5. Over the next few weeks, look for changes in your ecosystem. Twice each week add a pinch of fish food to the ecosystem. Add water as needed to maintain the level.

ANALYZE AND CONCLUDE

On separate paper, write your answers.

1. What function does the plant serve in this ecosystem?
2. What is the function of the snail?
3. What role do the goldfish play in the ecosystem?
4. Why is the light needed for your ecosystem?
5. What are you providing to the ecosystem on a regular basis? Why?

Multicultural Women of Science ◆ Ann Haven Morgan   *© 1996 The Peoples Publishing Group, Inc.*

think WORK ACT

CRITICAL THINKING Answer the following questions in complete sentences.

1. Why was the work done by Ann Haven Morgan important?

   _____

   _____

2. Why should an understanding of ecology be included in science education?

   _____

   _____

3. How does your concept of conservation tie in with the subject of ecology?

   _____

   _____

GOING FURTHER Complete three of the following.

BUILD YOUR PORTFOLIO

Survey your community for examples of organisms in their natural environment. Photograph, draw, or write a description of each organism you observe and its surroundings. Identify the living and nonliving parts of the organism's environment.

CONCEPT MAPPING

Use a dictionary to find the meanings of the terms *environment, habitat, niche, ecology, conservation, ecosystem,* and *natural resources.* Develop a concept map that shows how these terms are related.

JOURNAL WRITING

Prepare a radio or television script for an editorial that suggests ways people can work together to improve conditions in their environment.

COOPERATIVE LEARNING

Do research to find out what food chains and food webs are and how they are shown. Cut out pictures of organisms from magazines. Use the pictures to make several food chains and webs.

RESEARCH AND REPORT

Collect newspaper and magazine articles dealing with environmental issues. Write a summary that identifies the main concern in each article. Explain how people are addressing each concern.

DIAN FOSSEY
1932-1985

The gorilla is often portrayed as an ill-tempered, chest-beating beast. Although the gorilla is powerful and may beat its chest when excited or alarmed, this animal is actually gentle, shy, and curious. Much of what is known about the mountain gorilla has largely resulted from research done by Dian Fossey. For nearly twenty years, Fossey studied the habits and social structure of this endangered species.

Dian Fossey was born in San Francisco, California, in 1932. As a child, she was not allowed to keep any pets, other than goldfish. Still, she loved animals. This love of animals greatly influenced her adult life and eventually led to her death.

After completing high school, Dian Fossey attended the University of California at Davis. She enrolled in the preveterinary medicine program. Fossey later transferred to San Jose State College where she studied occupational therapy and received her degree in 1954. Two years later, she moved to Kentucky and became director of the occupational therapy department of the Kosair Crippled Children's Hospital in Louisville.

> ### Vocabulary
> **Primates** are mammals with flexible hands and feet, each with five digits, including humans and gorillas.

As a child, Fossey had always wanted to visit Africa. In 1963, her lifelong dream came true. Against the advice of her family and friends, she left her job and took out a bank loan to pay for her journey.

While in Africa, Fossey met with noted anthropologist Louis B. Leakey and his wife Mary at a dig site in Tanzania. Louis Leakey had earlier encouraged Jane Goodall in her study of chimpanzees. Leakey was impressed with Fossey's enthusiasm and determination to study the mountain gorilla. He became even more convinced of her commitment when, after breaking her ankle, she still made the long hike to the Democratic Republic of the Congo, now called Zaire, to observe gorillas.

In her book, *Gorillas in the Mist,* published in 1983, Fossey vividly described her first meeting with the mountain gorilla. She was overwhelmed by the sight, sound, smell, and physical size of the animals.

In spite of her wish to stay in Africa, Fossey returned to the United States to continue her work with physically-challenged children. In 1966, Louis Leakey met once again with Fossey. He convinced her to return to Africa to continue her studies of the mountain gorilla.

In 1967, Fossey set up her first observation camp in east central Africa. At first, she was able to observe the gorillas only from great distances; however, she later learned to imitate gorilla feeding and grooming habits and to copy

PRIMATOLOGIST

many of their sounds. These skills allowed Fossey to be accepted by the gorillas. She was then better able to observe the gorilla families up close.

Fossey studied the primate's systems of communication, eating habits, and nesting patterns. She was most impressed by the gorilla's strong family ties and how they cared for their young. She also conducted population studies of the species and worked hard to prevent humans and their activities from harming the gorillas. Over a period of nearly twenty years, Dian Fossey became a leading expert on the habits and social structure of the mountain gorilla.

Around 1970, Fossey left Africa to study for her doctorate at Cambridge University in England. In 1974, after receiving her degree, she returned to Africa. Failing health caused by severe dietary deficiencies forced her to return to the United States in 1980. At that time, she accepted a teaching position at Cornell University. At the end of her term at Cornell, Dr. Fossey returned to her research camp in Rwanda, Africa.

As Fossey continued her field studies, tensions grew between her and local villagers. The villagers felt strongly that land set aside as a gorilla refuge was badly needed for living space. In addition, Fossey waged a constant and dangerous campaign against poachers—illegal hunters—who killed the gorilla as a trophy, for its parts, or captured them for sale to foreign zoos. Tensions continued to mount. Fossey was slain in her camp, in December of 1985. Authorities in Africa blamed her death on area poachers.

PERSPECTIVES

At about the same time that Dian Fossey was preparing to leave for Africa to study the mountain gorilla, R. Allen Gardner and his wife Beatrice T. Gardner, psychologists at the University of Nevada, were attempting to communicate with chimpanzees. The Gardners knew that chimps often communicate using hand gestures and that the throat of the chimpanzee is very different from the human throat. Thus, they decided to communicate with the chimps using sign language rather than sound.

The ability of primates to communicate with each other and humans may someday permit researchers to learn more about these animals, especially their habits and social organization. It may also help scientists understand how these animals think and feel. With this knowledge, scientists may one day be able to explain many characteristics of the animal world that are today poorly understood.

It's **YOUR TURN**

Dian
FOSSEY

Hands-On Activity

OBSERVING ANIMAL BEHAVIOR THROUGH INDIRECT EVIDENCE

MATERIALS (per group of two)

owl pellet, dissecting needle, hand lens, forceps, protective gloves, face mask

SAFETY

Use the dissecting needle carefully to avoid injuring yourself. When handling owl pellets, wear a face mask and protective gloves.

BACKGROUND INFORMATION

Dian Fossey did her work in the field rather than in a laboratory. Scientists working in the field must often make indirect observations of animals and their behavior. By studying tracks, a scientist can determine whether a particular type of animal was in an area, when, and which direction it was heading. Such observations can provide many clues about a particular type of animal.

When an owl eats, it cannot digest all parts of its food. The undigested parts are stored in the owl's gizzard. An owl eliminates these parts in a ball-shaped mass called an owl pellet. By studying owl pellets, scientists can learn about the diet of owls.

PROCEDURE

1. Place the owl pellet on your desk or table top. Observe the owl pellet with and without the hand lens and describe its appearance in your notebook.

2. Use the dissecting needle and forceps to carefully break apart the owl pellet. Arrange the materials separated from the pellet on your desk in groups representing like parts.

3. Use the hand lens to observe the characteristics of the materials you pulled apart from the owl pellet. Describe the parts and where you think they may have come from in your notebook.

ANALYZE AND CONCLUDE

On separate paper, write your answers.

1. Describe the appearance of the owl pellet when it was first given to you.

2. After you pulled apart the owl pellet, what types of materials appeared to be present?

3. Based upon the appearance of the materials found in the pellet, do you think owls feed on plants, animals, or both? Explain.

4. What kinds of information can be learned about an owl by examining an owl pellet?

5. How might indirect evidence be helpful to a scientist?

think WORK ACT

CRITICAL THINKING Answer the following questions in complete sentences.

1. How did the work of Dian Fossey threaten the lives of local villagers and poachers?

   _____

   _____

2. Why would being able to imitate the sounds made by an animal be helpful to a person studying the animal?

   _____

   _____

3. How was Dian Fossey's study of mountain gorillas the fulfillment of a lifelong dream?

   _____

   _____

GOING FURTHER Complete three of the following.

BUILD YOUR PORTFOLIO

Survey your neighborhood looking for tracks left by five different kinds of animals. Make drawings or take photographs of the tracks you find. Try to identify the animals that left the tracks. Describe any other conclusions you can make about the animals based on the appearance of their tracks.

CONCEPT MAPPING

Use library resources to find out the differences among threatened, endangered, and extinct species. Make a concept map that explains these differences.

JOURNAL WRITING

In your journal, describe the difficulties you might face while conducting a study such as that done by Dian Fossey.

COOPERATIVE LEARNING

Conduct library research to find out which endangered animals live in your state. Make a table that identifies the animal, identifies why it is endangered, and explains what efforts are being made to save the species.

RESEARCH AND REPORT

Jane Goodall began field research to study chimpanzees a few years before Dian Fossey conducted her research. Research the work of Jane Goodall. Prepare an oral report on the contributions Goodall has made to science.

WOMEN OF

MARIA TELKES
1900-

CATHERINE
LITTLEFIELD GREEN
1755-1814

TECHNOLOGY

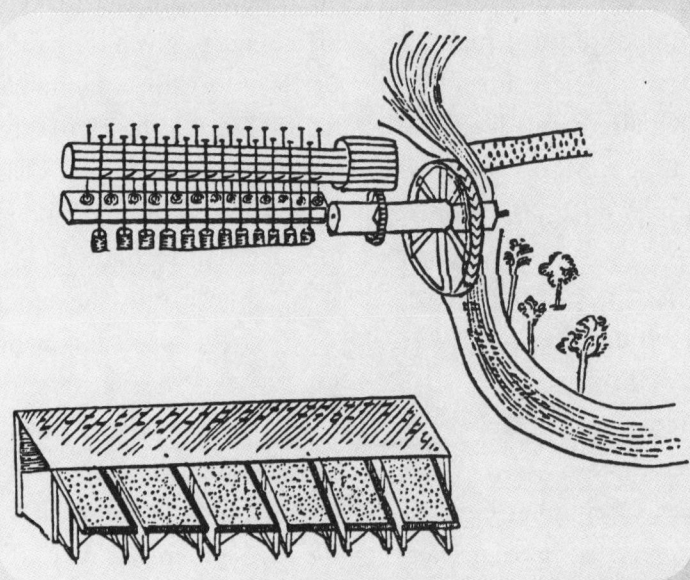

SYBILLA MASTERS
1680?-1720

The first patent issued to an American was British patent #401 for a machine to clean and cure Indian corn invented by Sybilla Masters, 1680?-1720. Patents were not issued to women, so the patent was in Thomas Masters' name, Sybillia's husband.

The Industrial Revolution in the early 1800s was a period of mechanical invention. It was a time marked by the use of machines. It was also a time when anything that could burn was used to supply energy for these machines. Fuels, such as wood, coal, and oil, were the fuels of choice. Today, these fuels are still in use and in greater amounts than ever before. Limited supplies of these fuels and pollution concerns suggest the need for other energy sources. One such alternative is solar energy, the energy of the sun. Maria Telkes was a pioneer in the use of solar energy as a source of heat and power.

Maria Telkes was born in Budapest, Hungary, in 1900. Telkes received her early education in a convent school where she showed a keen interest and talent for science. During her high school years, she grew interested in solar energy and its possible use as a power source. She read books and articles about solar energy that were written in English, French, German, and Hungarian.

Telkes attended the University of Budapest. In 1920, she earned her bachelor's degree. Four years later, she received her doctorate in physical chemistry. Physical chemistry is the science which studies the relationship between the physical properties and the chemical make-up of substances. While working on her doctorate, Telkes taught physics in Budapest.

In 1925, Telkes traveled to the United States to visit a relative and decided to stay. Her arrival in the United States marked the beginning of a brilliant career in the physical sciences. From 1926 to 1937, she worked as a biophysicist at the Cleveland Clinic Foundation. Her work there led to the invention of a sunlight powered meter that could record the electrical energy given off by the human brain.

Telkes became a United States citizen in 1937. During the 1940s, Telkes worked as a researcher in the use of solar energy at the Massachusetts Institute of Technology (MIT). During this time, she designed a solar heating plant for use in home heating. Earlier systems developed at MIT circulated water. A heating plant developed at the University of Denver used crushed rock to absorb and give off heat. Neither system was very efficient. Dr. Telkes created a different design. In her design, chemical crystals absorbed and stored solar energy. The crystals then slowly released the energy as heat. During the summer, the system maintained a comfortable temperature by removing hot air from the house. The system was efficient and inexpensive. Since it did not use fossil fuels, the system was also non-polluting.

> ### Vocabulary
> Solar energy is the energy given off by the sun.

In 1941, the United States entered World War II. Shortly afterwards, the U.S. government asked Telkes to act as advisor to the Office of Scientific Research and Development. One of her major research projects was to develop a water distillation system for use on life rafts. Her design used solar energy to change sea water into fresh water that humans could drink. In 1948, Telkes was called upon again by the government, this time to study the shortage of drinking water in the Virgin Islands. She changed her earlier distillation system, again using solar energy. For her efforts, Telkes was awarded the Certificate of Merit by the Office of Scientific Research and Development.

Telkes has published many articles. She also holds many patents dealing with solar heating and thermoelectricity. She has held research positions with Westinghouse, at the College of

APPLICATION OF THE SCIENCE

Solar energy can be used to produce heat, light, and electricity. Since the 1940s, scientists have done much research on how to use solar energy in different ways. For example, scientists use solar panels called vanes on space vehicles. These vanes provide power for the vehicle and the instruments on board the vehicle. An advantage to using solar energy for this purpose is that it reduces the need to carry great quantities of fuel and heavy batteries into space, reducing the weight and size of the space vehicle. Solar panels are also used to provide electricity for navigational lighting in remote waterways. In addition, scientists have experimented with solar-powered cars that are less polluting to the environment than gas-powered models.

It's **YOUR TURN**

Maria
TELKES

Hands-On Activity

COMPARING THE RATE AT WHICH DIFFERENT MATERIALS GIVE OFF HEAT ENERGY

MATERIALS (per group of four students)

two mayonnaise jars with holes in the lids, two laboratory thermometers, sand, water, two rubber bands, light source, graph paper, clock or watch

SAFETY

Wear safety goggles and a laboratory coat or apron throughout this activity. Secure the rubber bands and insert the thermometers through the holes in the lids carefully to avoid breaking them. If you do break a thermometer, notify your teacher immediately. Carefully fill jar with water. Wipe up spills immediately.

BACKGROUND INFORMATION

Many materials absorb light energy. Often, this energy is changed to heat energy that is given off by the object. The rate at which this process occurs varies for different materials and determines their usefulness as a heat source.

PROCEDURE

1. Half-fill one jar with water. Half-fill the second with sand. Screw a cover onto each jar.
2. Secure a rubber band around each thermometer about 10 cm above the bulb.
3. Insert the thermometer through the hole in the lid of the water-filled jar so it goes down into the water approximately 2 to 3 cm below the surface of the water. Move the rubber band to secure the thermometer in this position.
4. Repeat step 3 for the jar containing the sand.
5. Place both jars about 25 cm from the light source. Turn the light on and allow the jars to remain undisturbed for ten minutes.
6. Shut off the light and record the temperature in each jar. Continue recording the temperature for each jar at two-minute intervals for ten minutes.
7. Graph your results for both jars, plotting time versus temperature.

ANALYZE AND CONCLUDE

On separate paper, write your answers.

1. When the light was turned off, which container had the higher temperature?
2. Calculate the total heat loss for each substance during the ten-minute period after the light source was turned off. To make this calculation subtract the final temperature from the starting temperature.
3. Divide each number recorded in question 2 by ten to determine the heat loss per minute interval.
4. Which substance gave up its heat energy more slowly?
5. Of the materials tested, which would you use for a solar heating plant? Why?

think WORK ACT

CRITICAL THINKING Answer the following questions in complete sentences.

1. What are some advantages and disadvantages of using solar energy in place of fossil fuels?

_____

_____

2. Why is solar energy considered a renewable energy source?

_____

_____

3. How might the area in which one lives affect the use of solar heating plants for home heating?

_____

_____

GOING FURTHER Complete three of the following.

COMMUNITY RESOURCES

Contact your state or regional Environmental Protection Agency office to find out what energy sources are currently in use in your state. Ask for information regarding how these sources affect the environment.

CONCEPT MAPPING

Do research on the following energy sources to find out if they are renewable or nonrenewable and polluting or non-polluting. Consider these sources: oil, coal, natural gas, uranium, wood, corn, alcohol, wind, geothermal, solar, and moving water. Organize your findings in one or two concept maps.

JOURNAL WRITING

Describe and illustrate three ways you use solar energy in your daily life.

ALTERNATIVE ASSESSMENT

Design an experiment to determine how different concentrations of salt (NaCl) affect the rate at which water absorbs and gives off heat energy. Include at least three different concentrations and a control in your experiment. Submit your design to your teacher for approval before continuing. Once you have approval, carry out your experiment.

RESEARCH AND REPORT

Determine the annual home-heating expense for your home. Do research to find out how much it would cost to install a solar heating unit in your home and what the projected savings over time of installing such a unit are. Write a summary of your findings.

MARY ENGLE PENNINGTON
1872-1952

Chances are the milk you drink does not come from a local dairy. Your fruits and vegetables may come from thousands of miles away. What keeps these foods from spoiling before they reach your local supermarket? Many foods are kept fresh through refrigeration. Mary Engle Pennington was a pioneer in the use of refrigeration to keep foods fresh.

Mary Engle Pennington was born in Nashville, Tennessee, in 1872. The Pennington family later moved to Pennsylvania. As a child, Mary spent many hours in the garden with her father. She developed a strong interest in plants. Her interest in science was further sparked when she read a book on medical chemistry at age twelve. This subject greatly influenced the direction Mary's life would take.

In 1890, Mary went to the Towne Scientific School of the University of Pennsylvania. She studied chemistry and biology. In two years, she completed the requirements for a bachelor's degree; however, Mary was denied the degree because she was a woman. She was instead given a certificate of proficiency.

Mary was allowed to continue her studies at the University of Pennsylvania. Ironically, the same university which denied her a Bachelor's degree in 1890 because of gender permitted her to receive a Ph.D. in chemistry in 1895. She spent two more years at the university studying botany. She then spent one year at Yale studying

chemistry. Yet, after all these years of education, Mary still had not received her bachelor's degree.

Pennington returned to Philadelphia in 1898. At this time in history, it was hard for women to find work. To work in her field, Pennington opened the Philadelphia Laboratory Clinic, where she performed medical tests. Pennington developed an excellent reputation which led to her appointment as a lecturer at the Women's College of Medicine of Pennsylvania. She held this position until 1906.

Pennington sought government employment, but women were not generally given jobs with the government. On the advice of Dr. Harvey Wiley, Chief of the Bureau of Chemistry of the United States Department of Agriculture, Pennington took the tests needed for government employment as M.E. Pennington. The use of initials hid the fact that she was a woman. Pennington succeeded in her goal and became a bacterial chemist for the Department of Agriculture in 1907.

In 1908, Pennington became chief of the Food Research Laboratory. The laboratory studied ways to handle and store food. The general concern of the laboratory was to determine when food became spoiled; however, its main goal was to prevent spoilage from taking place. Much of Pennington's work at the laboratory resulted in food packaging, transportation, distribution methods and standards that are still in use.

> **Vocabulary**
>
> **Refrigeration** is a method of preserving food by lowering its temperature.

CHEMIST

Pennington did much work on the effect of humidity and freezing on foods. During World War II, she developed a design for refrigerated railroad cars that transported food throughout the country. The design won her a Notable Service Medal from the Hoover administration. In 1940, she received the American Chemical Society's Garvan Medal, an annual award honoring a woman chemist. Pennington's other honors include becoming the first woman elected to the Poultry Historical Society's Hall of Fame and the first woman member of the American Society of Refrigeration Engineers.

APPLICATION OF THE SCIENCE

Refrigeration and freezing are not the only methods used to preserve foods. Other methods for preserving foods include drying, salting, and smoking. Many of these methods date back to early civilizations. More recent methods of food preservation include the addition of chemical preservatives such as MSG or nitrates, canning, vacuum-packing, irradiation, and genetically altering plants and animals to be used as food. Genetically altered plants and animals tend to be more resistant to disease, larger, and better tasting than foods grown or raised using traditional methods.

Not all foods are preserved using the same methods. The method of food preservation best suited to a particular food is determined by how it affects the flavor and texture of the food and how long the food will be stored. Scientists continue to look for ways to prevent foods from spoiling. Because of space and weight requirements, the space program is likely to produce new methods of preserving and storing foods.

Multicultural Women of Science ◆ Mary Engle Pennington

It's **YOUR TURN**

Mary Engle
PENNINGTON

EFFECTS OF REFRIGERATION ON FOOD

MATERIALS (per group of 3)

whole milk, three baby food jars with lids, glass-marking pencil, refrigerator with freezer compartment (for class use)

SAFETY

Clean up any spills immediately. Never taste any materials in the laboratory.

BACKGROUND INFORMATION

Most foods spoil because of the action of bacteria. Refrigeration is a means of slowing down bacterial action. The lower the temperature, the more slowly the action occurs. Thus, refrigeration and freezing are effective methods of preserving foods.

PROCEDURE

1. Use the glass-marking pencil to label your jars *A*, *B*, and *C*. Identify each jar with your group number.
2. Pour equal amounts of milk into each jar. Cap each jar.
3. Place jar A in the freezer compartment of a refrigerator. Place jar B in the refrigerator. Place jar C in a cabinet or on a shelf in your classroom as provided by your teacher.
4. Check the milk in each container daily looking for any changes of color or odor which indicate that the milk has spoiled. To observe the milk's odor, hold the open jar away from your face, and use your hand to fan the odor of the milk toward your nose. When you observe that spoilage has occurred, record the jar letter and date in your notes. Discard the jar as instructed by your teacher.

5. Repeat step 4 for the other two jars. Throw away any remaining jars after one month.
6. Compare your results for steps 4 and 5 with those of the other groups in your class.

ANALYZE AND CONCLUDE

On separate paper, write your answers.

1. In which jar did you first observe signs of spoilage? Where was this jar stored? How long did it take for spoilage to occur?
2. How did your observations for the other two jars compare with the observations you noted in question 1?
3. How did your observations for all jars compare with those of other groups in your class?
4. What can you predict about the milk in your last remaining jar?

think WORK ACT

Mary Engle PENNINGTON

CRITICAL THINKING Answer the following questions in complete sentences.

1. Why is it important for some foods that are shipped long distances to be refrigerated?

_____

_____

2. What factors other than temperature might affect food freshness?

_____

_____

3. How do you think food prices are affected by the ways foods are preserved and transported?

_____

_____

GOING FURTHER Complete three of the following.

BUILD YOUR PORTFOLIO

Create a one-week menu for yourself and your family. Consider such factors as cost, storage, preparation time, and nutrition. Explain the food choices included in your menu.

ALTERNATIVE ASSESSMENT

Visit a supermarket looking for three different types of items that are packaged and stored in more than one way. Record the type of packaging and method of storage for each product as well as information about its shelf life and price. What conclusions can you make about how packaging and storage affect shelf life? How do these elements appear to affect price?

COOPERATIVE LEARNING

Do research to find out what irradiation is and how it is accomplished. Find out why some people think irradiation is a good idea while others are opposed to this practice. Organize a debate based upon your findings.

JOURNAL WRITING

Many fruits and vegetables can be purchased fresh, in cans, or frozen. In your journal, describe how the fruits and vegetables you buy are most often packaged. Explain your choice.

RESEARCH AND REPORT

Do library research about the methods of food preservation discussed on p. 151. Write a brief description of each process and the types of foods each is used to preserve.

CATHERINE GREENE
1755-1814

Cotton is one of the most common fibers used by people. Cotton fibers are harvested from bolls that form as the flowers of a cotton plant wither. After the cotton boll is picked, the fibers must be separated from the seeds. A machine called the cotton gin performs this task. Eli Whitney is generally credited as the inventor of the cotton gin; however, the idea for the cotton gin may have been developed by a woman named Catherine Littlefield Greene. Greene may also have perfected this device.

Catherine Littlefield was born in 1755 in a fishing and farming village on Block Island, Rhode Island. She was the third of five children. Her father represented the town in the colonial assembly.

Catherine attended the village school in New Shoreham. In 1774, she married Nathaniel Greene, a distant relative. Greene enlisted in the Revolutionary Army and became one of George Washington's most trusted generals. Catherine Greene accompanied her husband in his military travels. She endured, along with the soldiers, the bitter winters at such places as Valley Forge, Schuylkill, and Morristown. Three of her five children were born during these grim years.

> ### Vocabulary
> A **mixture** is when two or more substances are placed together but do not combine chemically.

When the Revolutionary War ended, the state of Georgia gave General Nathaniel Greene an estate to express its thanks to him for his military service. The family moved to Mulberry Grove, a plantation on the estate. Less than one year after the move, Nathaniel Greene died.

Following Nathaniel Greene's death, a Yale graduate named Phineas Miller took over management of the debt-ridden plantation. As a favor, Miller got a recent Yale graduate named Eli Whitney to act as a tutor to a neighboring family's children, however, the tutoring job did not work out. Catherine Greene then hired Whitney to develop a machine that could separate cotton seeds from cotton fibers.

Eli Whitney was given a basement room in which to work. He worked on the device for nearly six months. By April of 1793, Whitney had made a working model of the cotton gin. This model had wooden teeth that did not work very well. After studying the device, Catherine Greene suggested that Whitney use wire brushes in place of the wooden teeth. Whitney made this change.

The new model of the cotton gin was a success. But announcing the machine to the public before the device was patented was a mistake. Copies of the machine appeared everywhere. Neither Whitney, his partner, Phineas Miller, nor Greene received any money from the sale of

INVENTOR

these machines. Finally, on March 14, 1794, Eli Whitney was granted a patent for the cotton gin.

Catherine Greene began a legal battle to try to maintain Eli Whitney's rights to the cotton gin. The battle lasted for almost ten years. During this time, neither she, Miller, nor Whitney made any profit from the invention.

In 1796, Catherine Greene married Phineas Miller. Four years later, the couple was forced to move to a smaller plantation on an estate called Dungeness. In 1803, Miller died. Four years later, in 1807, the legal battles over the cotton gin ended. Although Eli Whitney was credited as the inventor of the device, his patent, which ended the same year was not renewed. Catherine Greene remained at Dungeness until her death at age fifty-nine.

PERSPECTIVES

A patent is a government grant that gives an inventor the right to keep others from making, using, or selling an invention without permission for as long as twenty years. This protection allows inventors to be the first to profit from their inventions.

The idea for a patent system in the United States began in 1787 with the U.S. Constitution. Three years later, the first patent law was signed by President George Washington. This law gave the job of reviewing and granting patents to a committee made up of the Secretary of State, the Secretary of War, and the U.S. Attorney General. Today, the issuing of patents and the registering of trademarks is the responsibility of the U.S. Patent and Trademark Office. This office is part of the Department of Commerce.

Catherine GREENE

It's YOUR TURN

Hands-On Activity

SEPARATING MIXTURES

MATERIALS (per group of two or three)

two mixtures provided by your teacher, bar magnet wrapped in a plastic bag, comb, strainer, filter paper, funnel made from top part of 2-L plastic bottle, plastic jar, four paper plates

SAFETY

Wear safety goggles and a laboratory coat or apron throughout this activity. Clean up any spills that occur immediately

BACKGROUND INFORMATION

A mixture is made when two or more substances are placed together but do not combine chemically. Examples of mixtures include air and ocean water. A cotton boll is also a mixture of cotton fibers and seeds. Because the substances in a mixture do not combine chemically, they can be separated from each other using physical methods.

PROCEDURE

1. Study the mixtures given to you by your teacher. As a group, decide how to use only the materials listed to separate the parts of the mixtures from each other. Write out your plan.
2. Have your plan approved by your teacher. After you get approval carry out your plan. Place each recovered material in a separate container (plate or jar).

ANALYZE AND CONCLUDE

On separate paper, write your answers.

1. What materials did your group use to separate mixture 1? Why?
2. Could you have separated the materials in mixture 1 in another way? How?
3. What materials did your group use to separate mixture 2? Why?
4. Could you have separated the materials in mixture 2 in another way? How?

think WORK ACT

CRITICAL THINKING Answer the following questions in complete sentences.

1. Why might Catherine Greene deserve at least part of the credit for being the inventor of the cotton gin?

   _____

   _____

2. Suggest reasons to explain why Eli Whitney is generally credited as being the sole inventor of the cotton gin.

   _____

   _____

3. What benefits do you think the cotton gin provided to farmers who made their living growing cotton?

   _____

   _____

GOING FURTHER Complete three of the following.

JOURNAL WRITING

Imagine you are Catherine Greene. Explain why you think you should be given some credit for the development of the cotton gin.

COMMUNITY RESOURCES

Write to the United States Patent and Trademark Office to find out how a person obtains a patent. Make an outline that summarizes the steps involved in the process. Include a copy of your letter and your outline in your portfolio.

PERFECT YOUR SKILL

Find examples of fifteen common trademarks in newspapers or magazines. Make a poster that identifies the company associated with each trademark.

COOPERATIVE LEARNING

A trademark is a word, name, slogan, design, or symbol used to identify the maker of a product. Work with your group to develop a product you would like to sell. Describe or make a model of the product, explain what the product does, and give it a name. Then choose a name and design a trademark for your company. Present your company and your product to the class.

RESEARCH AND REPORT

The making of the cotton gin marked the beginning of the Industrial Revolution in the United States. Find out how long this period of history lasted and what other major inventions were developed at this time. Organize your findings in a timeline.

Have you ever thought of inventing something? If you do invent something, you will need to protect your invention. In many countries, inventors protect their inventions by getting a patent. A patent gives the inventor all legal rights to the making, use, and sale of the invention. Obtaining a patent can be a time-consuming and costly process. Why, then, would someone patent their invention in another country under someone else's name? Strange as it may seem, that is exactly what Sybilla Masters did. Masters was an American woman who received several patents in England in her husband's name.

Sybilla Masters was the second of seven children born to William and Sarah Richton, a Quaker family living in the colony of West New Jersey. Sybilla's father was a seafarer and merchant. Little is known about her mother. Because the old records are incomplete, little is known about Sybilla's place or date of birth or early life.

In the mid-1690s, Sybilla married Thomas Masters, a wealthy Quaker merchant. The Masters made their home in Philadelphia, Pennsylvania. While Thomas worked in politics or at his business, Sybilla raised their four children. She also created several mechanical devices.

Vocabulary

An invention is the creation of a new way of doing things or a new device.

In 1712, Sybilla and Thomas traveled to London, England. The purpose of the trip was to obtain patents for two of Sybilla's inventions. During the early 1700s, a patent office had not yet been created in colonial America. Three years later, Thomas Masters was granted a patent for one of Sybilla's inventions, in Thomas's name. The patent was for a device used to clean and cure Indian corn. The device, which could be powered by a water wheel or horses, powdered dry corn kernels by stamping them. Shallow storage bins were used to cure and then dry the corn meal. Finally, the corn meal was packaged and sold as a cure for pulmonary tuberculosis, then called consumption.

Anticipating success, Thomas Masters purchased a mill for large-scale production of the corn product; however, sales were not very good and the mill soon closed.

A second patent was issued to Thomas Masters in 1716 for an invention of Sybilla's. This invention was for a new method of working and staining straw and for working the leaves of the palmetto tree. The braided and stained materials were then used to cover and adorn hats and bonnets. After the patent was issued, Sybilla obtained the exclusive rights to import the palmetto leaf into England from the West Indies. She then opened a shop in London where she sold hats, bonnets, baskets, and matting for chairs and other furniture that were made from straw and the leaves of the palmetto tree.

Several months after opening her business, Sybilla and Thomas returned to Philadelphia. Thomas Masters petitioned the provincial government of Pennsylvania for permission to have Sybilla's patents recorded and published in Pennsylvania. Thomas Masters was granted this permission on July 15, 1717. Three years later, Sybilla Masters, who may have been America's first woman inventor, died.

PERSPECTIVES

Many inventions of the 1700s and early 1800s were created by women. However, credit for these early inventions was either not given at all or given to men, as illustrated in the case of Sybilla Masters.

In the early history of the United States, most states had laws that prevented women from owning property in their own names. If the woman was married, any property she had was controlled by her husband. Since inventions were considered property, women could not get patents in their own names. Thus, credit for an invention made by a woman was often given to the woman's husband or to some other man.

The Patent Act of 1790 gave both women and men equal rights in obtaining patents; however, the laws of most states which forbid women to own property were still in place. It was not until 1809 that a woman obtained a patent in her own name in the United States.

1790 George Washington signs the nation's first patent law—the Patent Act of 1790.

1809 Mary Kies becomes the first woman to obtain a U.S. Patent for a method of "weaving straw with silk or thread."

1794 Eli Whitney receives a patent for the cotton gin, however, evidence suggests the idea for the device was developed by Catherine Littlefield Greene.

1876 George Hibbard admits to the patent council that his invention of a feather duster made from turkey feathers was his wife Susan's idea. The patent council reissued the earlier patent in Susan Hibbard's name.

1885 Sarah E. Goode becomes the first African American woman to receive a patent. The patent is issued for a folding cabinet bed.

1904 Margaret Knight receives a patent for a rotary engine.

1921 A patent for the "Tommy iron" is granted to Frederick Kern. Kern later confesses that Bertha Thomson created the "Tommy iron."

1942 Actress Hedy Lamar is issued a patent for a secret communication system developed for use during World War II.

1954 Gertrude Belle Elion receives a patent for a leukemia drug called mercaptopurine.

1957 Rachel Brown and Elizabeth Hazen are awarded a patent for an antifungal drug called Nystatin.

1981 Lila Beauchamp is issued a patent for her development of anti-viral compounds.

1991 Nobel Prize-winning scientist Gertrude Belle Elion is inducted into the National Inventors Hall of Fame. Elion holds more than forty patents.

1994 Rachel Brown and Elizabeth Hazen are inducted into the National Inventors Hall of Fame for their development of Nystatin, for which they received a patent in 1957.

It's YOUR TURN

Sybilla
MASTERS

Hands-On Activity

MAKING A BATTERY TESTER

MATERIALS (per individual)

one AA battery, two rubber bands, masking tape,
small light bulb, two wooden splints, two paper clips,
pencil, two pieces of aluminum foil

BACKGROUND INFORMATION

A battery stores chemical energy. The chemical energy in a battery can be
changed to electrical energy and used to light a bulb. To do this, electrical energy must be able to flow from the negative end of the battery, through the light
bulb and back to the positive end of the battery. A battery will produce electrical
energy from its stored chemical energy until the battery goes "dead." A battery
tester can tell you if a battery is still useful as an energy source or if it is dead.

PROCEDURE

1. Look over the materials you have been
   given. Plan a way to use only the materials
   provided to make a battery tester.
2. Set up your battery tester according to
   your plan. If your battery tester does not
   work, try another plan. After you have
   made a battery tester that works, make a
   sketch that shows all the parts of your
   completed battery tester. Label all the
   materials shown in your sketch.
3. Exchange batteries with another classmate.
   Use the tester you designed to test that
   battery. Record what happens. Repeat this
   step, exchanging batteries with a third
   classmate.

ANALYZE AND CONCLUDE

On separate paper, write your answers.

1. What materials did you use to make your
   battery tester? Why did you select these
   materials?
2. Did your battery tester work the first time
   you tried it? If not, why not? If so, why?
3. Study your sketch. Suggest two materials
   that might be used in place of the materials you used in your tester. Explain your
   choices.

think WORK ACT

CRITICAL THINKING Answer the following questions in complete sentences.

1. Why were Sybilla Masters' inventions patented in her husband's name rather than her own name?

   _____

   _____

2. Why might it be better for Sybilla Masters to have had her inventions patented in her husband's name, than not to have had the inventions patented at all?

   _____

   _____

3. Why do you think the Masters had Sybilla's inventions patented in England before they had them patented in the United States?

   _____

   _____

GOING FURTHER Complete three of the following.

JOURNAL WRITING

If you made an invention, what would you want your invention to do? Write a brief response to this question in your journal.

ALTERNATIVE ASSESSMENT

Develop a system of measurement for length. Assign at least three units representing different lengths to your system. Explain how your system works and how the units are related. Describe the advantages and disadvantages of your measuring system compared with the measuring system you normally use.

COOPERATIVE LEARNING

Have each member of the group find information about ten patents issued to women. For each patent, find the year it was issued, what it was for, and the name of the woman who received it. You may want to have one member of your group write to the U.S. Patent and Trademark Office to get information. Combine the data gathered by your group in a timeline.

RESEARCH AND REPORT

Choose one event or invention from the timeline on p. 159. Conduct research on the topic and write a brief report explaining its importance.

Graph Grid

Date _____

Name _____

Barbara McClintock	BAHR bruh muh KLIN tock
Rosalind Franklin	RAHZ uh lund FRANGK lin
Mary Leakey	MARE ee LEE kee
Lynn Margulis	LIN MAHR guh lis
Alice Eastwood	AL iss EEST wud
Margaret Mead	MAHR guh rit MEED
Flossie Wong-Staal	FLAH see WONG-STAHL
Sara Josephine Baker	SERR uh JOH suh feen BAY kur
Jane Wright	JAYN RYT
Myra Adele Logan	MY ruh uh DELL LOH gun
Rita Levi-Montalcini	REE tuh LEE vee-MAHN tul SEE nee
Gerty Cori	GUR tee KOHR ee
Gertrude Elion	GUR trood EL ee un
Joycelyn Elders	JOS uh lin EL durz
Susan LaFlesche Picotte	SOO zun luh FLESH pih KOHT
Winifred Goldring	WIN ih fred GOLD ring
Florence Bascom	FLOHR uns BAS kum
Sylvia Earle Mead	SIL vee uh URL MEED
Joanne Malkus Simpson	JOH an SIMP sun
Ellen Ochoa	EL un oh CHOH uh
Henrietta Swan Leavitt	hen ree ET uh SWAHN LEV it
Annie Jump Cannon	AN ee JUMP KAHN un
Mae Jemison	MAY JEM ih sun
Marie Curie	muh REE KYOO ree
Irene Joliot-Curie	eye REEN ZHOH lee oh-KYOO ree
Lise Meitner	LEE zuh MYT nur
Maria Goeppert-Mayer	muh REE uh GUR payrt-MAY ur
Dorothy Hodgkin	DOR uh thee HOJ kin
Rosalyn Yalow	RAHZ uh lin YAL oh
Rachel Carson	RAY chul KAHR sun
Ynes Mexia	EE nez meh HEE uh
Ann Haven Morgan	AN HAY vun MOR gun
Dian Fossey	dy AN FAHS see
Maria Telkes	muh REE uh TEL kes
Mary Engle Pennington	MAHR ee EN gul PEN ing tun
Catherine Littlefield Greene	KATH rin LIT ul feeld GREEN
Sybilla Masters	suh BIL uh MAS turs

Glossary

AIDS acronym for acquired immune deficiency syndrome, a group of illnesses that result from a viral attack on the immune system

angiosperms flowering plants; plants that reproduce by forming seeds that are enclosed within a fruit

anthropology the study of the origin and development of humans

antibiotics substances that inhibit the growth of or kill disease-causing bacteria

arboretum botanical garden devoted to trees and shrubs

archaeology science that studies humans of the past based on information excavated from former dwelling sites or examination of remains

astronomy science that studies the stars, planets, and other bodies in space

B

bacteria single-celled organisms that do not have true nuclei or membrane-bound organelles

biochemistry the branch of biology that studies the chemistry of living things

biology branch of science that deals with the study of living things

biopsy surgical removal of a small amount of living tissue for microscopic examination

botany branch of biology that deals with the study of plants

C

cancer the rapid, abnormal continuous increase in the number of cells in a part of the body

carbohydrate a sugar or starch

chemistry branch of science that deals with the study of matter and its properties

chloroplast the cell structure that contains chlorophyll; usually found in plants

chemotherapy the use of chemical substances to treat disease

chromosomes tiny structures in the nuclei of living cells that carry the genetic information that determines the traits of an organism

clones genetically identical offspring to the parent

conservation the wise use of natural resources so that they are not used up and will be available in the future

constellation star pattern that resembles the shapes of people, animals, or other objects

cranial capacity a measure of the size of the part of the skull that holds the brain

crystallographer scientist who studies the traits of crystals

D

diabetes disease in which the inability to produce insulin in proper amounts results in an inability to properly control sugar levels in the body

dicot plant that produces seeds having two seed leaves

DNA acronym for deoxyribonucleic acid, the nucleic acid that controls the traits of organisms

E

ecology branch of science that deals with the study of the relationships between organisms and their surroundings

ecosystem all the living and nonliving parts of an organism's environment and their interactions

element a substance that cannot be broken down chemically

embryo developing organism between the stages of fertilization and birth or hatching

embryology the study of the development of organisms prior to their birth or hatching

endocrinologist scientist who studies the functions of the ductless glands in the body

ethnology branch of anthropology that deals with the studies of cultures

evolution process by which living things change over time

extinct any species that is no longer in existence

F

fossil any physical or trace evidence indicating the presence of an organism that lived in the past

G

genes parts of the chromosomes that carry the hereditary information

genetics branch of science that deals with the study of heredity, or how traits are passed from parent to offspring

genus a group of several species with many similarities

geology science that deals with the study of the structure, composition, and history of the earth, moon, and other planets

glucose sugar that is broken down by most organisms for the release of energy

glycogen one of the forms in which glucose is stored in the body

H

habitat the home of an organism

herbarium organized arrangement of dried plants

histologist scientist who studies the tissues or organisms, or prepares tissue for study

HIV acronym for human immunodeficiency virus, the virus that causes AIDS

hominid first humanlike organism to appear on Earth between 4 and 8 million years ago

I

immune system body system that helps protect the body from disease

insulin hormone produced by the pancreas which controls the use of sugar by the body

isotope form of an element that has the usual number of protons but a different number of neutrons from other isotopes of the same element

magnitude in astronomy, measure of brightness of a star

metabolism sum of all the chemical processes that take place in an organism

meteorologist scientist who studies weather and conditions of the atmosphere

mitochondria cell organelles that carry out respiration and provide energy to the cell

mixture two or more substances that are placed together but do not combine chemically

molecular biology the science that deals with the study of the structure and function of the chemicals in living things

molecular structure the exact atomic makeup of a molecule of substance

monocot plant that produces seeds that have only one seed leaf

mutation a spontaneous change in the genes or chromosomes of an organism

N

NASA acronym for National Aeronautics and Space Administration; U.S. government agency that deals with the exploration of space

natural selection theory of evolution proposed by Charles Darwin which states that organisms with traits adapted to their environments will survive and reproduce, while organisms not adapted to their environments will not survive and reproduce

nerve growth factor (NGF) chemical in nerve cells that controls the growth of these cells

NOAA acronym for National Oceanographic and Atmospheric Administration; U.S. government agency concerned with studies related to the ocean and weather

nova star that rapidly increases in brightness and then gradually fades

nucleic acids large, complex molecules that contain hereditary information

nuclear reactor a device in which nuclear fission occurs to produce energy for use by people

nucleus in biology the control center of a cell; in chemistry the central part of an atom in which the protons and neutrons are located

O

oncologist scientist or physician who studies cancer

organelles cell structures within the cytoplasm that act like small organs by carrying out specific jobs within a cell

P

paleontologist scientist who studies fossils

patent government grant to an individual or organization that gives an exclusive right for a certain period of time to make, use, or sell a new invention

pediatrician medical doctor who specializes in providing care to children

petrology branch of geology that deals with the study of the makeup and formation of rocks

photovoltaic cell device that changes the energy of the sun into electrical energy

physics branch of science concerned with the relationship between energy and matter

physiology science that deals with the study of the functions and processes that take place in living things

primates mammals with flexible hands and feet, each with five digits, including humans and gorillas

R

radiation process in which energy or particles are given off by matter

radioactivity the giving off of particles and energy from the nucleus of an atom

radioimmunoassay use of radioactive elements to trace or detect chemical processes occurring in the body

radioisotope radioactive isotope

replication the process by which DNA and viruses make copies of themselves

reservation a tract of public land set aside for Native Americans

retrovirus a virus that replicates in a manner that is opposite that of most viruses

S

seed bank a storage and collection site for seeds and plant parts from which new plants can be grown

seeding clouds injecting clouds with chemicals to induce rain

solar energy energy given off by the sun

spectra the band of colors that result when white light is passed through a prism or diffraction grating

spectroscope device that uses a prism or diffraction grating to separate light into its component colors

spectroscopy the study of the light given off by an object using a spectroscope

symbiosis a relationship in which two different organisms live and are dependent on each other for existence

T

tumors abnormal growth in body tissue

V

variable stars stars that vary in brightness at different periods of time

W

wavelength the measure of the distance from the crest of a wave to its trough

Z

zoology branch of the biology of animals

More Biographical Sketches of Notable Multicultural Women of Science

Adamson, Joy (1910 - 1980)
Joy Adamson is an Austrian artist who went to Kenya, Africa when she was 26 and remained there to work as a wildlife photographer. Her photography has earned her the Greenfell Gold Medal of the Royal Horticultural Society. Her most notable accomplishment came with the writing of four books, *Born Free, Living Free, Elsa: The Story of a Lioness,* and *Forever Free,* which detail the life, death, and offspring of a lioness named Elsa. Three of these books, which were made into movies during the 1960s, helped focus worldwide attention on wildlife conservation. Using monies earned from the sale of the books Adamson worked to establish the Elsa Appeal, a foundation devoted to the preservation of wildlife.
Suggested Readings: *Born Free, Living Free, Forever Free*

Agnodice (~300 BCE)
Agnodice was a Greek physician of the third centur BCE. She is credited with helping to change Greek law to allow free women living in the city-state of Athens to practice medicine. Before Agnodice became a physician, the practice of medicine was illegal for both free and enslaved women. To get around this law, Agnodice, attended medical school and practiced her craft disguised as a man. After her identity was discovered, she was imprisoned and tried for breaking the law. Pressure from her women patients led to a change in the law. Free women were permitted to practice medicine, but were restricted to treatment of only women patients.

Anderson, Elda Emma (1899 - 1961)
Anderson was a health physicist who served as a member of the team that developed the atomic bomb during World War II. Anderson earned her bachelors degree from Ripon College in 1922, her masters in physics from the University of Wisconsin in 1924, and her Ph.D. in physics from Wisconsin in 1941. After her work on the development of the atomic bomb, Anderson became an expert in health physics, concentrating her efforts in protection from radiation.

Anderson, Elizabeth Garrett (1836 - 1917)
Elizabeth Garrett Anderson was the first English woman to be permitted to practice medicine in Britain. After being refused admission to medical school, she studied on her own. In 1865, she was licensed to practice medicine by the Society of Apothecaries in London. Other "firsts" in her life included being the first woman member of the British Medical Association and the first woman mayor in Britain. Anderson was a cofounder of the London School of Medicine and served as a physician at the Maryleborne Dispensary for Women and Children, a hospital which was later renamed for her.

Aspacia (~2nd century CE)
Aspacia was a physician who specialized in obstetrics, gynecology, and surgery. She lived in the Roman empire during the 2nd century CE. Much of what is known about the treatments and surgeries performed by this early woman doctor has been reported through the writings of people living several centuries later. She was most known for her use of herbs in the treatment of women experiencing difficult pregnancies and in her advice to pregnant woman to eat lightly and avoid strenuous exercise.

Blackwell, Elizabeth (1821 - 1910)
Elizabeth Blackwell is generally recognized as the first woman physician of the United States. Blackwell was the first woman to be granted her medical degree from the Geneva Medical School. She was responsible for the establishment of the New York Infirmary for Indigent Women and Children. Her writings include *The Physical Education of Girls,* published in 1852 and *Pioneer Work in Opening the Medical Profession to Women,* which was published in 1895.

Blunt, Katharine (1876 - 1954)
Blunt worked as a college administrator, educator, home economist, and nutritionist. She received her bachelors degree from Vassar College in 1898 and continued her education at the Massachusetts Institute of Technology. For a period of time, Blunt worked for the United States Department of Agriculture and the Food Administration, specializing in the development of foods for use by soldiers during wartime. In 1925, Blunt became chair of the home economics department of the University of Chicago. In 1929, she became the third president of Connecticut College for Women, a position she held until her retirement in 1943.

Braun, Emma Lucy (1889 - 1971)
Emma Lucy Braun was a botanist and conservationist. She is noted as one of the pioneer ecologists of the first half of the twentieth century. During her life, she published more than 180 papers and books on topics related to plants and gardening. Her combined degrees in geology and botany led her to devote much of her life to the study of physiographic ecology. Among her many accomplishments are the organization of the Ohio Flora Committee in 1951 and the founding of the Cincinnati chapter of the Wild Flower Preservation Society. Braun also served as the first woman president of the Ohio Academy of Science and the Ecological Society of America.

Cobb, Jewel Plummer (1924 -)
Jewel Plummer Cobb is a contemporary African American scientist who studies the effects of anticancer drugs on human cancer cells. In the past, Cobb has worked with noted African American cancer researcher Jane Cooke Wright of the Harlem Hospital in New York to study how melanin helps protect skin from ultraviolet radiation.

Cobb, Isabelle (1858 - 1947)
Isabelle Cobb was the daughter of European American Joseph B. Cobb and Native American Evaline Clingan Cobb who was of Cherokee decent. In her youth, Isabelle Cobb was educated in the public schools of Cleveland, Ohio and at the Cherokee Female Seminary in Ohio. Cobb later attended the Women's Medical College of Pennsylvania, from which she received her medical degree in 1892. After receiving her MD, Cobb went to Oklahoma to work with Native Americans. Because the area she worked in was very rural and far from a hospital, Cobb was often forced to perform surgical procedures in the homes of her patients.

Collins, Eileen (1956 -)
Eileen Collins is a NASA astronaut who was selected for the space program in 1990. In 1994, Collins made history when she became the first American woman to pilot a shuttle mission. Collins holds a BA in mathematics from Syracuse University; a master of science degree in operations research from Stanford University, and a masters degree in space systems management from Webster University.

Gray, Dr. Ida (1867 - 1953)
Ida Gray was the first African American woman to earn of Doctor of Dental Surgery degree. Born in 1867, Gray attended the Gaines public school in Cincinnati, Ohio. She later attended the University of Michigan Dental School from which she received her DDS degree. As testimony to her achievements, the epitaph on Gray's grave marker reads: "Dr. Ida Gray Nelson Rollins, First Negro Woman Dentist in America."

Harris, Mary Styles (1949 -)
Mary Styles Harris is a contemporary African American geneticist who specializes in the study and treatment of sickle-cell anemia. Harris served as the executive director of the Sickle-Cell Foundation of Georgia from 1977 - 1979. In 1980, *Glamour* magazine named her as one of their Outstanding Women Scientists.

Hunt, Fern Y. (1948 -)
Fern Hunt is a contemporary African American applied mathematician who holds a master's degree and a doctorate in mathematics from New York University. Hunt did her undergraduate work at Bryn Mawr College.

While at Bryn Mawr, Hunt read about the work of mathematician Sonya Kovalesky, whose work inspired Hunt to earn her degree in mathematics. Hunt has served as a professor of mathematics at Howard University.

Hypatia (370 - 415 CE)

Hypatia was an African mathematician and inventor from Egypt. She received her education from her father, who was a professor at the University of Alexandria in Egypt. After completing her education, Hypatia returned to the university as a teacher. As an inventor, Hypatia is often credited with the development of the astrolab and the planesphere, devices used to study stars. She also invented the hydroscope, an instrument used to measure the specific gravity of water. Hypatia became a victim of political and religious conflict for her beliefs. Under the direction of Cyril, the ruler of Egypt, Hypatia was dragged from her chariot by a mob and tortured to death.

Jackson, Shirley Ann (1946 -)

Shirley Ann Jackson is a native of Washington, D.C. who became the first African American in the United States to receive a doctorate in physics from the Massachusetts Institute of Technology (MIT). Jackson earned her degree in 1973 and later went to work as a research assistant at the Fermi National Acceleration Laboratory. In 1982, she served as a lecturer at the NATO International Advanced Study Institute in Antwerp, Belgium and later went to work for the AT&T Bell Laboratories in New Jersey.

Kenny, Sister Elizabeth (1886 - 1952)

Sister Elizabeth Kenny was a nurse from Australia who developed a method for treating people suffering from poliomyelitis, also called infantile paralysis. Prior to the mid-1950s, this disease plagued millions of people worldwide. She wrote of her experiences in the treatment of poliomyelitis in her autobiography, *And They Shall Walk,* which was published in 1943. Other works written by Kenny include *Physical Medicine Concerning the Disease Infantile Paralysis* (1945) and *My Battle and Victory*

(1955). For many years, the Infantile Paralysis Association made use of Kenny's medical practices. They abandoned Kenny's methods in search of a cure for poliomyelitis. Sister Kenny received her nursing degree in 1902 and practiced her profession in the bush country areas of Queensland, Australia.

Kies, Mary (? - ?)

Mary Kies is the first woman to ever be granted a U.S. patent. Kies, who was a British citizen, obtained her patent in 1809 for her process of weaving straw using either silk or thread. At the time, the U.S. government had stopped importing European goods in response to the Napoleonic Wars. As a result of the process developed by Kies, the hat-making industry in the United States prospered.

Maathai, Wangari (1940 -)

Wangari Maathai is a Kenyan environmentalist who during the 1970s became popular throughout Africa for her Green Belt Movement. The movement, which began in the 1970s, is responsible for the planting of millions of trees throughout the African continent. In addition to her work as an environmentalist, for which she received the Right Livelihood Foundation's Goldman Environmental Prize in 1985, Maathai has achieved many firsts in her life. Among these achievements is her standing as the first woman in eastern and central Africa to hold an advanced degree, the first woman to serve as the head of a university department in Kenya, and the first to become an assistant professor at the University of Nairobi, Kenya.

Mitchell, Maria (1818 - 1889)

Maria Mitchell is recognized as the first professional woman astronomer of the United States. Mitchell, who was self-taught, began her training in astronomy while working as a librarian in Nantucket's Athenaeum. During her working hours, Mitchell read books on astronomy, as well as other sciences and mathematics. At night, she used her father's telescope to study the sky. She made her first significant astronomical contribution in 1847, when she discovered a comet. The following year, Mitchell became the first woman to be elected into the American Academy of Arts and Sciences. Despite her lack of a college degree, Mitchell went on to serve as a professor at Vassar College. Later, she helped found the Association for the Advancement of Women. Following her death, an observatory was built on Nantucket Island in her honor.

Ocampo, Adriana C. (1955 -)

Adriana Ocampo is a Latin American planetary geologist who was born in Barranquilla, Columbia in 1955. During at least part of her career, Ocampo has worked for the National Aeronautics and Space Agency (NASA). While with NASA Ocampo worked on project Galileo, an exploratory mission of the planet Jupiter.

Taylor, Lucy Hobbs (1833 - 1910)

Lucy Hobbs Taylor became the first African American woman in the world to earn a degree in dentistry. Taylor earned her degree from the Ohio Dental School in 1866.

Trotula (11th century CE)

Trotula gained notoriety in the 11th century as one of the greatest medical women of Italy. Trotula lived at a time when Europe was devastated by plagues and male doctors were in short supply. These factors enabled Trotula and other women to study medicine. Trotula received her medical training and later worked as a teacher at the Salerno, the earliest non-religious school in Europe. Working with her husband and son, both of whom were physicians, Trotula was involved in the creation of an encyclopedia of medicine called *Practica Brevis*. Trotula later compiled a text devoted to gynecology and obstetrics, titled *De Mulierum Passionibus* (On the Suffering of Women). The book which was originally prepared in Latin editions that were hand copied is the earliest known text devoted to women's health care. After mechanical printing was developed, the book was printed in quantity and was produced in several languages and many reprintings for almost 300 years.

Wauneka, Annie Dodge (1910 -)

A Navajo born in Sawmill, Arizona, Wauneka became a noted public health activist and politician. Throughout her youth, Wauneka became fluent in both Navajo and English and served as a translator to both groups. Wauneka earned a BS in public health from the University of Arizona. Both before and after receiving her degree, Wauneka focused much of her attention on improving the health of her people. She increased awareness about the importance of innoculations against certain diseases and improved sanitation. In 1959, she received the Arizona State Public Health Association's Outstanding Worker in Public Health Award and the Indian Achievement Award from the Indian Council Fire of Chicago. Wauneka later served as a member of the Surgeon General's Advisory Board and as a member of the National Tuberculosis Association. In 1963, she received the Presidential Medal of Freedom from President John F. Kennedy.

Suggested Readings

If you are interested in learning more about the work of the scientists included in this book, this partial listing of books and articles will get you started.

Rachel Carson
Rachel Carson, Marty Jazer, Chelsea House Publications, New York, 1988.
Newsweek, Vol. 38:86., July, 16, 1951.

Gertrude Elion
New York Times Magazine, "The Nobel Pair," Katherine Bouton, January 29, 1989.
Washington Post, "Pathway to the Prize," Don Colburn, October 25, 1988.

Mae C. Jemison
Ebony, Vol. 44:50, 52, 54-55, August, 1989.
Essence, Vol. 19:59-60, October, 1988.

Marie Curie
Madame Curie, Eve Curie, Da Capo Press, New York, 1986.
Grand Obsession, Marie Curie and Her World, Rosalynd Pflaum, Doubleday, New York, 1989.

Mary Leakey
Disclosing the Past, Mary Leakey, Doubleday, New York, 1984.
National Geographic, "The Leakeys of Africa: Family in Search of Prehistoric Man," February, 1985.

Barbara McClintock
Women Inventors and Their Discoveries, Ethlie Ann Vare and Greg Ptacek, Oliver Press, Minneapolis, Minnesota, 1993.
A Passion to Know, Allen L. Hammond, Charles Scribner and Sons, New York, 1984.

Margaret Mead
Margaret Mead, A Portrait, Edward Rice, Harper & Row Publishers, New York, 1979.

Ynes Mexia
Woman in the Field, Marcia Myers Bonta, Texas A&M University Press, 1991.

Ann Haven Morgan
Woman in the Field, Marcia Myers Bonta, Texas A&M University Press, 1991.

Mary Pennington
Mothers of Invention, Ethlie Ann Vare and Greg Ptacek, William Morrow, New York, 1988.

Jane Cooke Wright
Women Pioneers of Science, Louis Haber, Harcourt, Brace, Jovanovich, New York, 1979.

Flossie Wong-Staal
Contemporary Women Scientists, Lisa Yount, Facts on File, New York, 1994.

Rosalyn Yalow
Mothers of Invention, Ethlie Ann Vare and Greg Ptacek, William Morrow, New York, 1988.

Index

H

I

J

K

L

P

pH, p. 123
paleontology, p. 12
parasite, p. 19
Parkinson's disease, p. 47
Parkman, Paul, p. 55
parsec, p. 89
patent, pp. 84, 155, 158, 159
pediatric endogrinology, p. 58
penicillin, p. 118
Pennington, Mary Engle, p. 150
Pennington, Mary Engle, Awards and honors,
  lecturer at Women's College of Medicine of
  Pennsylvania, p. 150; bacterial chemist for Dept.
  Of Agriculture, p. 150; chief, Food Research
  Laboratory, p. 150; Notable Service Medal, p.
  151; American Chemical Society's Garvan Medal,
  p. 151; first woman elected to Poultry Historical
  Society's Hall of Fame, p. 151; first woman
  member, American Society of Refrigeration
  Engineers, p. 151
Period Luminosity Relationship for Variable Stars, p. 89
pesticides, p. 129
PET scan, p. 39
petrology, pp. 68, 72
physical chemistry, p. 146
physics, physicist, p. 103, 106, 146; nuclear
  physics, nuclear physicist, p. 111; quantum
  mechanics, p. 114
Pickering, Edward, p. 88
Picotte, Susan LaFlesche, p. 62
Picotte, Susan LaFlesche, Honors and awards, first
  Native American woman to receive medical
  degree, p. 62; Chair of local Board of Health, p.
  63; founded hospital on Omaha reservation, p.
  63; organized County Medical Society, p. 63
Planck, Max, p. 110
poliomyelitis, p. 55
polonium, p. 102
pressure, water, p. 78
primates, p. 140
proteins, p. 122
protons, p. 114
public health, p. 58

Q

quantum mechanics, p. 114
quinine, p. 63

R

radium, p. 102
radioactivity, p. 102, 110; discoveries in, p. 103, 110
radioimmunoassay (RIA), p. 122
radioisotope, p. 122
rain, p. 8; rain forest, p. 132
refrigeration engineering, p. 151
reservation, p. 62
retrovirus, p. 28
RIA, *See* radioimmunoassay
Rutherford, Earnest, p. 115

S

Sabin, Albert, p. 55
Sagan, Carl, p. 16
Sagan, Dorian, p. 17
Salk, Jonas, p. 55
scuba diving, p. 76
seed banks, p. 21
seeding clouds, p. 80
shamans, p. 133
Sierra Club, p. 132
Simpson, Joanne Malkus, p. 80
Simpson, Joanne Malkus, Awards and honors,
  meteorologist, Woods Hole Oceanographic Institute,
  p. 80; Professor of Meteorology at University of CA,
  LA, p. 80; consultant for U.S. Weather Bureau, p. 81;
  director, Experimental Meteorological Lab of NOAA,
  p. 81; named 1963 Woman of the Year by *LA Times*, p. 81
smallpox, p. 55
solar system, p. 89
solar energy, p. 147
solar vanes, p. 147
solute, p. 120
solution, p. 120
spectroscopy, p. 92; spectroscope, p. 94
spectrum, p. 93
spinoffs, p. 97
spirilla, p. 32
stains, p. 47
stars, pp. 88, 92; variable, p. 88; red and blue-
  white, p. 88; see also *nova*
surgeon general, p. 58
symbiosis, p. 16

T

taxol, p. 132
Telkes, Maria, p. 146
Telkes, Maria, Awards and honors, Certificate of
  Merit from Office of Scientific Research and
  Development, p. 146; Advisor to Office of
  Scientific Research and Development, p. 147;

Multicultural Women of Science